gene(s), 21, 22-24, 34, 37, 40,
45, 49, 83, 86, 87
geneticist(s), 34, 100, 101, 104,
105
glume(s), 7, 29, 30, 40, 41, 44,
45, 46, 48, 67, 74, 92
grass(es), 7, 18, 28
Gray, Asa, 7, 9, 40

Harshberger, John, 15, 22
heredity, 19-24, 37
Hopi Indians, 19
hybrid(s), (see also corn, hy-
brid), 15, 43
inbreeding, 34
Indian(s), 50, 57, 69, 74
Aztec, 12
Hopi, 19
Ojibway, 10-11
Peruvian, 46
Tehuacán Valley, 94-99
Zuni, 30

Jones, Donald F., 105

knobs, chromosome, 100-101

La Perra Cave, 71-75, 89
Libby, Willard, 70

McClintock, Barbara, 100-101
MacNeish, Richard, 70, 71-75,
88-99
maize, see corn
origin of name, 50

corn (continued)
pod corn, 7-9, 23, 24, 26, 40-
42, 44, 45-49, 57, 62, 66,
83, 85-87
pollen, 78-82
pollination, 25, 27, 34
popcorn, 47, 48, 49, 66, 67,
73, 85
pre-Columbian distribution,
56-60
tassel-seed, 44
wild, 3-9, 18, 26-32, 44-49,
62, 80, 83-87, 88-99, 101
wild, reconstructed, 26-31, 83-
87
corn products industry, 106-108
Correns, Karl, 20
coyote corn, 14, 15
Cutler, Hugh C., ix-x

Darwin, Charles, 3-9, 10, 19, 24,
26, 34, 40, 76
de Bonafous, Mathieu, 54
de Candolle, Alphonse, see Candolle
Devil's Canyon, 71-75
De Vries, Hugo, see Vries, Hugo
De
Dick, Herbert, 64, 65-70

East, Edward Murray, 37, 99,
104, 109

fossil(s), 75, 76-82

Galinat, Walton C., 66, 67, 70,
72, 74, 84-87, 94, 101

The mysterious grain

THE

MYSTERIOUS GRAIN

by Mary Elting and Michael Folsom

illustrated by Frank Cieciorka

Published by

M. EVANS AND COMPANY, INC.,

New York

and distributed in association with

J. B. LIPPINCOTT COMPANY,

Philadelphia and New York

FOR DANK

Contents

Foreword   ix

The missing ancestor   3

Promises and disappointments   10

Invisible messages   19

Wild corn reconstructed   25

Just collecting evidence   34

Some pieces fall into place   43

The where of it all   50

Theory and practice   61

Evidence in Bat Cave   65

Devil's Canyon corn   71

Fossils   76

Wild corn reconstructed again   83

Combined operation   88

Where next?   100

Acknowledgments   111

For further reading   113

Index   116

Foreword

Here is a book which combines mystery, adventure and biography: the search for ancestors of a valuable plant, the trials of scientists in the wilds of Mexico, and the day to day planning of investigations. Rarely does a writer have a subject like this, which ranges from problems of primitive man to the latest research for improving quality, yields and utilization of our most useful crop.

The kinds of plants a man uses tell of his past. Man domesticated plants he found growing about his home and carried them with him during his migrations. Over the years the plants changed as man selected them for special purposes. These changes record his progress and travels. Wheat was carried out of western Asia and the Mediterranean region. Rye is associated with central Europe. Watermelons and sorghum spread from Africa. Rice speaks of contacts with southeastern Asia. Corn belongs to the Americas and has helped to make its people "healthy, wealthy, and wise" enough to use well a new food from the American Indian. Most of the grain we eat is wheat, and this reflects the habits and past of our ancestors. But the wealth of the Americas is based largely on Indian corn.

Although few of us grow our own foods, we are still closely bound to plants. In 1830 four persons were supplied with farm products by one farmworker. In 1930, before hybrid corn, improved methods of planting and fertilizing, and large pieces of mechanized equipment were in wide use—about ten people were provided for by a single farmworker. Now nearly forty people are supplied with farm products by one farmworker. Some of the people living off the farm spend their time building the machines and processing the fuel farmers use, but the increase in how much a single farmer can produce is still spectacular.

Everyone who reads this book will discover how closely linked are man and his plants and how they have evolved together.

Hugh C. Cutler
Curator of Useful Plants
Missouri Botanical Garden

The mysterious grain

The missing ancestor

When Charles Darwin was a young man he spent five years sailing around the world, studying plants and animals, collecting specimens by the thousands. He spent the rest of his life at home in England working on a theory that would explain the wonders he had seen. Darwin knew as much about nature as anyone in his time, and his theory of evolution did help men to understand how the countless varieties of living things had come to develop on earth. But he was bothered by one thing he had never found and one problem he could not solve.

In all his travels Darwin had never come across a wild corn plant. What corn's ancestors looked like he could not imagine. Nor could he think of a way to explain how or where cultivated corn developed.

This mysterious plant, which Darwin like all Englishmen called maize, comes in more shapes and sizes and colors than does any of man's other grains. Some kinds of corn have many ears, some only one. Certain ears that Darwin measured were four times as long as others.

There were narrow ears and thick ones, pointed cobs and cobs that branched. An ear might have only six rows of kernels or more than twenty. The kernels of one kind are so small that it takes nearly 16,000 of them to weigh a pound. In other varieties there are only about three hundred kernels to the pound. Darwin had seen kernels in many colors—yellow, white, orange, red—and a few, he observed, were elegantly streaked with black. Often an ear had kernels of several colors and shapes next to each other on the cob.

Such a host of differences fascinated Darwin because he had been the first to say clearly and precisely what they meant: Variety in living things was the result of evolution. But corn posed a special problem. It seemed to defy the obvious rule that plants can evolve only if they are able to survive. Among all the varieties of corn Darwin had seen, not one was able to survive by itself. Without man's help it could not scatter seeds to produce new plants year after year.

An ear of corn is put together in an odd tight bundle. The close-packed rows of kernels, which are the seeds, lie anchored securely in place underneath layers of husks. The ear itself grows firmly fixed to the stalk. There is no way for the seeds to be dispersed unless man peels the husks away, then forces the kernels off the cob, and puts them in the soil.

Darwin could imagine what would happen if a field of corn was left unattended. During the autumn and winter many of the stalks would topple over and ears would reach the ground. If they weren't eaten by animals or destroyed by mold, hundreds of seeds would begin to thrust up new

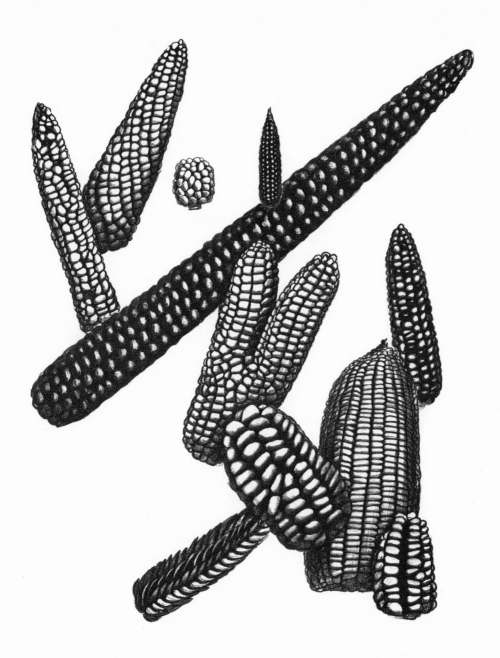

plants next spring—all in one small area around the cob. Each plant would struggle for water, sun, and nourishment from the soil. Like a horde of starving men competing for a crust of bread, all but a few of the new corn shoots would die. Most of those which survived would be too weak to produce new ears and new seeds. In a year—or two or three at the most—no sign of corn would remain in the field.

How could such a plant have evolved? Darwin knew that men had not always been farmers. So, obviously, corn must once have been a different plant which *could* scatter seed and grow wild. His interest was aroused even more because corn seemed not only helpless but also unique. There are other domesticated grain plants—wheat for instance —which also have changed under cultivation and have become dependent upon man. But botanists have found wild varieties of wheat which can be compared to field wheat. It is not difficult to imagine what the ancestors of modern wheat were like and how the plant evolved under the care of man.

Corn, however, seemed completely cut off from its past. Darwin had found no variety that could even give him any hints about the plant's origin and evolution. He had seen no wild grain that might resemble an ancestor of modern domesticated corn. Perhaps the wild plant was extinct. Or perhaps it might still be growing in some corner of the earth he had never visited. Since his own explorations had not led him to the missing ancestor, he turned to others for help.

Darwin read reports of explorers, and he corresponded with other scientists to find out what they knew. One promising lead appeared in a book on geographical botany by a Frenchman, Alphonse de Candolle, who had summarized almost all that was known about the maize plant's history.

What interested Darwin most was Candolle's report of still another Frenchman's discovery—a peculiar variety of corn which was supposed to grow in the jungles of Brazil. This odd specimen had a double set of wrappings around its seeds. Not only was the ear enclosed in the usual husks, but also each kernel on the cob grew inside a small, papery covering of its own. (Now this variety is commonly called "pod corn" because of the pod-like jackets around the kernels.)

Darwin wondered if the pods might be a trait which wild corn had. Most other members of corn's large family—the grasses—do have just such individual wrappings around each seed. The little husks, which botanists call *glumes,* protect the seeds. Because of its long glumes, pod corn looked more like other grasses than did modern corn. It seemed more primitive, less evolved in its own special direction.

Was pod corn the wild ancestor?

Darwin wanted to think so, but he needed more information. So he consulted Asa Gray, an American botanist who was one of the first to give full-hearted support to Darwin's new ideas about evolution. Gray had actually been experimenting with pod corn and had discovered something strange. Its

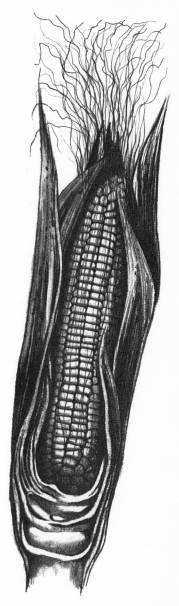

*A nineteenth-century
French drawing
of pod corn
studied by Darwin
and Candolle*

seeds often grew into two different kinds of corn. Some pod-corn seeds produced pod-corn plants, but other seeds from the same batch produced plants with normal, naked seeds.

In one generation this unusual plant seemed to change and become a quite different kind.

It was tempting to think that the change might be evolution going on before men's very eyes. But Darwin was sure that living things evolved in slow, small steps, not in such big jumps. There had to be some other explanation for the quick jump from pod corn to naked corn—an explanation which neither he nor Gray could figure out.

Reluctantly Darwin decided that pod corn could not be ancestral wild corn. Like other varieties, it had kernels that were fixed tightly to the cob and covered by outer husks. It could not sow its own seeds or survive outside man's fields and gardens.

Darwin had to leave the problem unsolved. Later scientists found reason to dismiss pod corn entirely in the search for the wild ancestor. But some came back to look at it in different ways and to pry new information from it. In the meantime, Darwin had started something. Even before his death in 1882, other men were busy using his ideas in searching for the origin of the maize plant.

Promises and disappointments

Long before Charles Darwin developed his theory of evolution and began to puzzle over the origin of corn, the Ojibway Indians were sure they knew the whole story.

In the days before men became farmers, the Ojibways said, a young man once complained that hunting and fishing were hard; he and his people did not have enough to eat. Not long afterward, he came upon a tall handsome stranger, all dressed in green, with a plume of dancing feathers over his pale, silky hair.

The stranger challenged the youth to wrestle with him. They wrestled all that day and all the next. Finally, after six days, the stranger (who turned out to be Mondamin, the corn god) admitted that he was beaten and said he would soon die. As a reward for the young man's strength and courage, Mondamin offered to help the Ojibway people, provided his instructions were followed.

Accordingly, as he was told to do, the young man buried the stranger in a soft grave and tended it with care. After a while he discovered growing from the grave a tall handsome plant, the likes of which he had never seen—with green leaves and pale, silky

10

threads like hair, and a feathery plume of tassels on top.

From that day on, the Ojibways did not lack food, so long as they took good care of the corn god's gift.

This Ojibway story was only one of many such old Indian legends. Almost every corn-growing tribe took an interest in the origin of the maize plant. Some told tales about gods or heroes or witches who had given the first corn to their ancestors. Some said that the gods had created man and woman out of corn itself.

In their poetic way, these legends told much that was true. The process of conquering nature to get food *is* a wrestle. The farming Indians *were* in a sense created out of corn, which was their staff of life. (In some Indian languages the word for corn is "that-which-gives-us-life.") But, like all the known varieties of corn, the legends had one thing in common: The plant they told about was essentially the same plant we know today. There were no stories about a wild plant that had been found growing without man's help.

Ancient Mexican pictures of corn-growing. At each stage a god protects the crop from such dangers as animals and drought.

Evidently the tales were invented after all memory of wild corn had been lost. In fact, they reflected the very mystery that botanists wanted to solve. The Indian myth-makers seemed to have realized that corn was the special possession of men. Unlike berries or deer, it could not be found growing wild. The Indians could only imagine that it was a miraculous gift.

*Teosinte with
detail of seed spike*

The stories of modern tribes gave little help to anyone who wanted to investigate the origin of the maize plant. However, Alphonse de Candolle, after he gathered all he could learn about the origin of corn, speculated that the Indians *did* in fact hold the clue to the mystery. Someday, Candolle guessed, archeologists would dig into the past and find among the relics of ancient Indian tribes real evidence of what corn's wild ancestor had been like.

In the meantime, botanists found they could investigate corn's past by examining its present. They turned to the plant itself and to its living relatives for information. And corn did have relatives. It was not so alone as it seemed. Two plants, botanists decided, were the maize plant's close kin.

The Aztec Indians had known about one of them for centuries. Corn they called *cintl* (SIN-tul). This other plant—a weed which grows in Mexico and Central America—they called *teocintl,* or "sacred corn." When the Spanish explorers came along, they pronounced and spelled the word *teosinte* (tay-oh-SIN-tay), and that is what we call it today.

Teosinte is a grass that looks something like corn. Its stalks and leaves are thinner, and each plant has a number of stalks growing out of a common root. Like corn, teosinte has two separate flowering parts: At the top of each stalk is the male part, a branching tassel which bears pollen. Farther down, at joints of the stalk, are bundles of small leaves enclosing the female flowers, and there the seeds grow. These bundles somewhat resemble an ear of corn. But inside a teosinte "ear" there is no cob—just a small spike bearing six or eight seeds.

When a teosinte "ear" is ripe, the surrounding leaves loosen, the brittle seed spike breaks, and the seeds scatter. Teosinte grows wild. Among all the grasses, it turns out to be most nearly like corn.

After teosinte had been identified as a relative of corn, botanists recognized still another—a plant called Tripsacum. (Since there is no popular name for Tripsacum, we use the Latin term which botanists have adopted.) Tripsacum scarcely resembles corn at all. It grows in clumps of thin, long leaves out of which spring slender stalks. The tip of each stalk is a kind of tassel, but each arm of the tassel bears both the male flowers and the seeds.

Tripsacum has nothing like the separate seed-bearing ear of corn lower down on its stalk. However, the seed spikes of teosinte and Tripsacum are very similar, and quite different from those of any other grass. The two plants are obviously related. Since corn is related to teosinte, it must also be a relative of Tripsacum.

Now botanists had something new to work with—new pieces of the puzzle to fit into place, new opportunities for speculation and experimentation. How did these three relatives form a family tree? What did teosinte and Tripsacum tell about the origin of corn?

In 1875 a German botanist named Paul Ascher-

Tripsacum

son came up with an appealing solution to the main problem. If he was right, no one could ever discover wild corn—because it never existed. The wild ancestor was a different species of plant altogether, and that plant, he said, was teosinte. The idea that corn evolved from teosinte seemed to be an easy way out of the mystery, and many people accepted it. But, before Ascherson's theory was tested, another even more promising candidate for the missing ancestor was discovered in Mexico. It had tiny ears. It grew wild. Mexican farmers said it was the ancestor of their cultivated maize, and they called it "coyote corn" because the kernels were hard and sharp-pointed like a coyote's teeth.

Coyote corn

In 1888 a number of these small wild ears reached Sereno Watson, a botanist at Harvard University. He planted the seeds and watched their growth with considerable enthusiasm. The ripened cobs broke easily into small sections, so that the plant could indeed sow itself without trouble. This seemed to be wild maize at last.

But when Watson made a detailed comparison of coyote corn and cultivated corn, he began to change his mind. Finally he decided that the two plants belonged to two different species.

Later it became clear that coyote corn was neither the wild ancestor nor a separate species. It was the product of a cross between corn and teosinte. In Mexico teosinte often grows adjacent to corn fields. The tassels of both plants shed pollen, which blows about freely, and the two plants are so closely related that the pollen of one can fertilize the other. Seeds that result from this cross-fertilization can grow into new plants which have some characteristics of both parents. (A cross between two varieties or strains of plant is called a *hybrid*.) In the case of the hybrid coyote corn, the ear did resemble a corn ear; like teosinte it was able to disperse its seeds. But this did not make it ancestral wild corn by any means.

The plant continued to get attention, however. Another botanist, John Harshberger, also called it wild corn. He later explained the mistake and settled the coyote-corn question for good. Meantime, his thinking about this wild hybrid plant had led him to a different theory about corn's origin. Modern corn, he said, did not evolve all by itself as an independent species. Rather it was a hybrid, the result of a cross between two different wild plants. He speculated that teosinte might have been one of the parents and some unknown wild grass the other.

This suggestion inspired more guesses. Other botanists made other suggestions about what two plants might have crossed to produce corn. Perhaps the maize plant had not one missing ancestor but two! The trouble was that no one could point to solid proof of hybrid origin, and for the time being many botanists dismissed the idea that hybrids played any part in the origin and evolution of modern corn.

Mayan corn god

The Aztec corn god, Cinteotl

Around the turn of the twentieth century the most usual theory was that corn had evolved directly from teosinte. This idea appealed to Luther Burbank. Here was something that he could test, he thought, and so he did.

Burbank was the most famous plant experimenter in America. Although he had not been trained as a botanist, he knew a great deal about how to breed flowers and vegetables and fruits, and he spent his long life searching and sniffing for oddities among the plants in his gardens. With much ingenuity and luck he selected the ones he wanted and bred them again and again, until he had new, more useful, or more pleasant varieties. Easter lilies, for example, did not have a strong, heavy fragrance when he began cultivating them. But he discovered "freaks" that did have a perfume, and from them he developed sweet-scented varieties. By careful selecting and breeding he created better potatoes, Shasta daisies, stoneless plums, and many other new varieties of plants.

In 1903 Burbank set out to use his breeding techniques on teosinte. He obtained seeds from Mexico and planted them on his experimental farm in California. At harvest time he selected seeds from the plants that showed the most corn-like traits. He grew these seeds and at the next harvest again selected the most corn-like descendants of his original teosinte seeds. After eighteen plantings and selections, Burbank had something that looked very much like

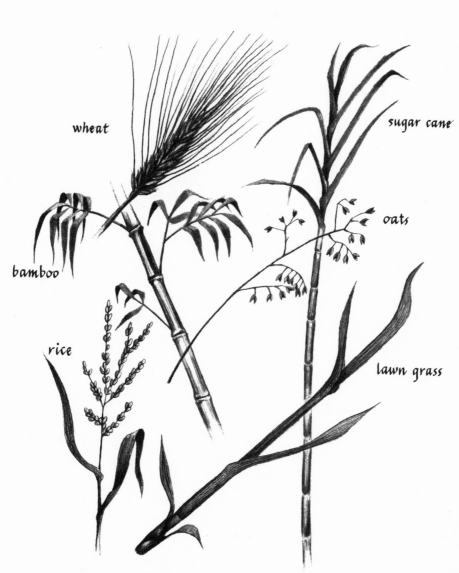

wheat

sugar cane

bamboo

oats

rice

lawn grass

Some members of the grass family

modern corn, with husk-covered cobs, quite unable to scatter seed and grow wild. He announced he had proved that teosinte was the wild ancestor of corn.

But scientists were skeptical about Burbank's results. First of all, they didn't believe that such a tremendous evolutionary change—from teosinte seed spike to corn cob—could take place in only eighteen generations. One botanist, Paul Weatherwax, who had been making close studies of every aspect of corn and its relatives, compared the seed spike of teosinte with the ear of corn and came to the conclusion that the one simply could not evolve into the other. The change would not have been possible, no matter how many generations an experimenter practiced selective breeding as Burbank had done.

The ancestor of corn, Weatherwax reasoned, must have been more primitive than corn, more like other grasses, less evolved in special ways. But teosinte looks as though it is highly specialized. In its own way it is as different from other grasses as corn itself is.

Moreover, when carefully controlled experiments were made, botanists could not produce from teosinte parents any offspring with the slightest tendency to resemble corn. It seemed impossible for teosinte to have been the ancestor of maize.

But Burbank had *proved* that teosinte could change into corn. What had really happened?

Invisible messages

Luther Burbank's teosinte experiment did seem to follow the pattern of evolution. First came the wild weed with its little spike of seeds; then, eighteen generations later, an ear of corn. It was certainly tempting to think this was what Darwin had called "the mystery of mysteries"—the appearance of a new species. Gradually the experimental plants had lost certain teosinte traits and developed certain traits of corn.

Darwin had shown that the same kind of thing happened in the natural process of evolution. But he had also asked himself again and again *how* it happened. How are traits passed along from parents to offspring? Why are traits lost, and where do new ones come from? He could only say that it seemed to him a seed must contain some invisible substance which determines the character of the next generation—like a message written in invisible ink.

The Hopi Indians thought about corn seeds in just those same terms. When they put red seeds in one row and yellow seeds in another, they were sending messages to the corn god asking for more of each kind. If at harvest time they found red kernels

Hopi corn-growing

19

on the same cob with yellow kernels, they simply assumed that the god had got the messages mixed up.

To late-nineteenth-century botanists, however, this phenomenon was a constant source of wonder. They knew that red kernels could develop if a plant in the yellow-corn row were fertilized by pollen from a red-corn plant. But *how* was a color trait handed down from parent to offspring?

Men who were curious about heredity did innumerable experiments with different varieties of corn, as well as with other plants, and early in the year 1900 three European botanists published reports outlining almost identical answers to the old puzzling questions. At the same time each of the three discovered that a fourth—an Austrian monk named Gregor Mendel—had come to similar conclusions about heredity more than thirty years earlier. Hugo De Vries in Holland, Karl Correns in Germany, and Erich Tschermak-Seysenegg in Austria had for the most part repeated Mendel's research, though each had contributed certain new details.

All four men had been intrigued by the fact that a plant or animal gets distinct traits from each of its parents, and that for each trait there seems to be what De Vries called a separate "unit" which passes from one generation to the next.

They figured out that in corn, for example, every male pollen grain in the tassel and every female egg cell on the cob contains a complete set of units—

one for height, one for color in the outside covering of the seed, one for the number of ears, and so on. When the pollen fertilizes the cell, the two sets come together, and there begins a kind of shuffling process. Presently a complete new combination of units is formed, made up of some from one parent and some from the other. This combination will determine what traits the offspring is to develop. Color of the seed may come from one parent, while the other parent may contribute the units for size and shape.

Naturally there arose a question about the units that have been left out of the new combination. Do they simply disappear?

The answer had to be no, although neither Mendel nor the other three botanists had actually been able to see the tiny units of heredity with a microscope. Whatever these minute particles might look like, they did not vanish. They remained hidden away in the cells of the next generation. And still hidden, they could even be passed along time after time. When conditions were right, a hidden trait might reappear—even hundreds of generations later.

As scientists went on searching they learned to decipher more and more of the "invisible writing" in the cells. They discovered that special threadlike bodies, called *chromosomes,* are made up of smaller units which do, indeed, control heredity. The units came to be called *genes,* and the study of them is *genetics.*

With a knowledge of genes and how they behave,

The Zapotec corn goddess, Centiocihuatl

it was possible for the first time to explain something so ordinary that most people didn't even wonder about it; that is, why a child may resemble a grandparent more than he resembles either his father or mother. The genes responsible for the development of head size or nose shape, for example, may have been hidden in the cells of his parents, only to reassert themselves in his.

Hidden genes could also explain the results of Luther Burbank's teosinte experiment. This is what must have happened: Burbank used seeds that were not pure teosinte. They came instead from a plant that had corn somewhere among its ancestors. As John Harshberger discovered, it often happens in Mexico that corn crosses with teosinte, and, if such a hybrid then crosses again with teosinte, the result can be a plant that looks very much like pure teosinte, although corn genes are there, hidden away.

At harvest time, Burbank examined his plants with a quick, practised eye. He had trained himself to pick out any that differed from the average, and among the "teosinte" crop he detected one or more that somehow resembled the corn ancestor. Burbank used seeds from these plants for his next sowing, and at harvest he again spied out the plants with the most cornlike traits. So it went, generation after generation. By careful selection, Burbank was able to get plants in which more and more characteristics of the corn ancestor were assembled. At the same time, the plants that showed traits of the teosinte ancestor were eliminated.

A corn-teosinte hybrid of the kind Luther Burbank used in his experiments

When botanists set out to test Burbank's results for themselves, they began with pure teosinte seed, and they made sure nothing but teosinte pollinated their plants, generation after generation. In the end, their test plants were no more maize-like than the first ones had been. With no hidden corn genes in its cells, their teosinte showed no slow, gradual tendency to evolve into corn.

The slowness of evolution was something most nineteenth-century scientists did not question. They had observed that individuals of any species differed slightly from one another, and when these small differences had been passed along to offspring, time after time, a plant or animal at last would have so many traits unlike its ancestors' that it could be called a member of a new species.

This idea of evolution by slow change did not entirely satisfy Hugo De Vries. He came to believe that living things could evolve in sudden, big steps. These dramatic changes he called *mutations*. A brand new trait would appear if there were a change in the actual makeup of a gene. The alteration was permanent, and so the new trait could be handed along from parent to offspring.

Was it possible that a mutation had turned teosinte into corn?

And what about pod corn? Were the shaggy individual wrappings around its kernels the result of a mutation that suddenly occurred in modern plants? Hugo De Vries did not think so. He said the pods must be the result of a hidden gene—a characteristic

Pod corn

of some ancient ancestor that turned up now and then. Like Darwin he half wished he could say that pod corn was true wild corn, but he decided he didn't have enough proof.

A great deal of searching, and some guesswork, had failed to track down the wild ancestor of the maize plant. During the first third of our century the study of genes would prove to be more and more important in the search. In those same years botanists also examined every other aspect of modern corn and its relatives; and their painstaking, cautious studies yielded promising clues.

Wild corn reconstructed

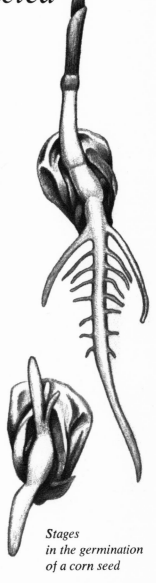

In 1923 Paul Weatherwax published a book called *The Story of the Maize Plant*. Like Candolle many years earlier, he had gathered up most of what was known about corn's origin and history. Candolle took only a few pages to tell what he knew; Weatherwax took many chapters. With deep interest and with vigorous contempt for what he considered nonsense, he recounted the studies, experiments, discoveries, false leads, and foolish guesses of the past. And he told in great detail what he and other botanists had learned from close scrutiny of corn and its relatives.

Even as a young man Weatherwax had a conservative temperament. He disliked wild speculation, and he was not the kind to get hold of an untried theory and run out to test it with an experiment. Facts interested him most. He discovered and collected a great many of them. Then he looked for the least complicated theory to explain them.

Every minute aspect of the corn plant came to his attention. He studied the structure of the tassel, and how its parts developed, opened, and scattered pollen in the wind. Under a microscope he watched what happened when pollen alighted on a moist silk.

*Stages
in the germination
of a corn seed*

25

He saw it begin to poke out a little tube which thrust its way down into the silk. Inside the grains of pollen he detected the tiny male sexual cells, the *sperm*. He probed deeper with his microscope and watched how the sperm were carried in the tip of the pollen tube down through the silk to the part on the cob where a new seed would appear. He made diagrams of the way in which the sperm and the female cell, the *ovule,* united. He noted how the seed developed, then what happened when a seed was planted, how it germinated, and how a whole new plant grew.

Weatherwax studied odd varieties of corn and "freaks" as well as the more usual kinds. There were ears that branched, cobs that had more rows of kernels at the bottom than at the tip, cobs that didn't seem to have any regular rows but just a hodge-podge of kernels all over. He looked at pod corn— the kind with the shaggy cobs which had interested Darwin and also another kind which grew pod-covered seeds at the base of the tassel.

While Weatherwax was trying to achieve a systematic understanding of the nature of modern corn, he was wondering about the old problem: Where had it come from? What was its history? In the end he put together all he and other botanists had learned about corn and evolution and came up not only with a theory about its origin, but also with a picture of what its wild ancestor must have looked like:

1   Neither teosinte nor Tripsacum could possibly be corn's direct ancestor;

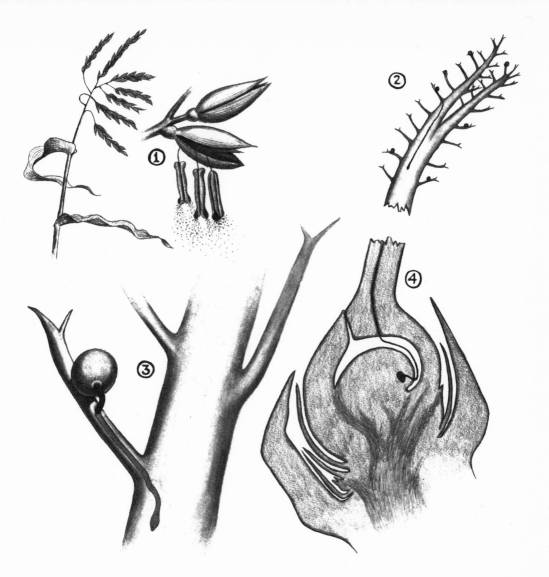

1. Pollen sacs (enlarged) in the tassel at the top of a corn plant burst open, and the pollen scatters. 2. (greatly enlarged) Pollen grains are caught on tiny hairs growing out of a corn silk. 3. (very greatly enlarged) From a pollen grain a tube emerges, and its tip pushes into the hair, then down through the silk toward the sac on the cob where a kernel of corn will develop. 4. (greatly enlarged) Pollen tube enters sac. The tube from a pollen grain can push through a silk more than a foot long. If the grain were enlarged to the size of a baseball, then its tube would travel a distance of several hundred feet.

*How the seeds of wheat,
Tripsacum, and corn
are attached*

2 there was no need to imagine any crossing between teosinte or Tripsacum and any other
grasses to account for the origin of corn;
3 *all three plants had, in fact, evolved independently from one common ancestor.*

This theory, Weatherwax argued, fitted all the
known facts and had the additional virtue of being
"reasonable, orthodox, and simple."

But how to determine what that lost common ancestor looked like? Weatherwax saw evidence that
the three relatives were once similar in all important
ways. For instance, one big difference between
modern corn and its cousins lies in the numbering of
their seeds. Corn kernels almost always grow in
pairs. That is why a corn cob has an even number of
rows. The seeds of teosinte and Tripsacum, however, grow singly. But, Weatherwax discovered,
tucked back under the single seeds on the spikes of
those two plants are little nubs—surely the rudi-
ments of second seeds which no longer develop. The
ancestor of all three plants probably grew seeds in
pairs.

In all three plants the seeds are deeply embedded
in the part which holds them. The seeds of teosinte
and Tripsacum snuggle into what look like ledges in
the spike. They can scatter because the whole spike
becomes brittle and breaks apart. Corn kernels are
anchored firmly in the cob. In other grasses the
seeds have various kinds of breakable attachment to
the spike. Probably the common ancestor of corn

and its relatives grew seeds, like other grasses, which were not embedded in their spikes.

Or take the glumes, the little leafy coverings which surround and protect the seeds of most grasses. Although the glumes *seem* to be lost in corn and its relatives, they aren't altogether missing. In teosinte and Tripsacum they have evolved into hard shells around the exposed portion of the seeds. In corn, what remains of the glumes is the chaff of the cob—the little papery ridges which are left when the kernels are chewed or shucked off the cob. With husks surrounding the modern cob, corn seeds have protection although they lack the individual wrappings. But the ancestor of the three plants surely had no husks, and probably it had papery coverings for each exposed seed.

The most remarkable and special thing about corn is its large ear. There is nothing like it anywhere else in the plant kingdom. But if you examine a cross section of an ear it is less than mysterious. The ear is a branch—collapsed like an accordion. Its structure is like that of the stalk, with thick, fibrous segments and leaves growing out of each joint, alternately on one side, then on the other. This branch, just like the stalk, ends in a spike—the cob—which bears flowers.

Unlike Tripsacum, corn and teosinte have two separate flowering parts—the pollen-bearing tassel at the top of the stalk and the seed spike or cob jutting out lower down the stalk. But Weatherwax

glumes

The glumes of corn and Tripsacum

saw varieties of corn which grew seeds at the base of the tassel and others which grew tassel parts at the tip of the cob. (The Zuni Indians call them "laughing ears" because they look funny.)

So, the ancestor of all three plants probably looked like this:

1   It had stalks with branches;
2   the branches ended in tufts of flowers which had both male and female parts;
3   the seeds grew in pairs, loosely attached to the spikes, and surrounded by protective papery glumes.

This imaginary ancestor was, of course, not wild corn. It was the ancestor of all three relatives—of corn and teosinte and Tripsacum. Wild corn must have evolved and existed as a separate plant before the first farmers began to cultivate it and before it evolved further into modern corn. How did corn evolve from this common ancestor in its own special direction?

Using what he knew of the plant's structure, Weatherwax imagined a number of changes happening slowly but more or less simultaneously:

1   The side branches became shorter:
2   the flowering parts tended to become wholly male or wholly female, with fewer kernels growing at the top of the stalk and fewer pollen-bearing parts at the ends of the branches;
3   more and more pairs of seeds began to grow out of the seed spike and the spike grew fatter and

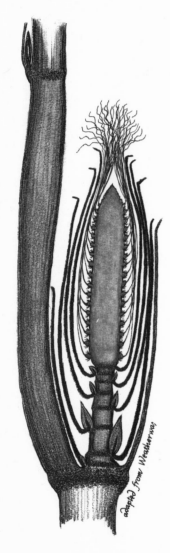

Cross section of an ear of corn

tougher—it began to look like a cob, with the seeds snuggling into its thick fiber, instead of standing free;

4   the glumes grew shorter;

5   the leaves on the branches, which once spread outward like those on the stalk, began to look like husks, hugging the cob and protecting the seeds;

6   the silks grew longer as the husks grew tighter around the cob.

All these changes made it harder and harder for corn to survive as a wild plant. Weatherwax made a guess to explain its survival. Wild corn, he said, may have been a perennial plant. That is, unlike modern corn, the same plant would live for many seasons, instead of dying each autumn and growing anew from its seeds the next spring. Many varieties of Tripsacum, and some varieties of teosinte, are perennials, and probably the common ancestor of all three plants was perennial too.

It also occurred to him that corn may have had the good fortune to be found at just the right moment. Perhaps its clumsy ears were already driving it toward extinction when human beings came upon it. The seed-packed cobs would have made it attractive to hungry men, and they may very well have saved it by learning to cultivate it in the nick of time.

Paul Weatherwax's reconstruction of the ancestry and evolution of corn was imaginative and useful. It had its base in real knowledge of many facts. As a

*"Laughing ears"—
a corn tassel
bearing seeds,
and a cob sprouting
a tassel part*

result of this kind of patient research and cautious reasoning, botanists had their first glimpse of the nature of wild corn.

But, no matter how accurate Weatherwax's reconstruction might be, there were still problems. It *was,* after all, guesswork. Weatherwax lacked evidence other than what he found by studying the modern plants. And he had not discovered what the searchers most wanted to find—the real thing: actual wild corn.

Cautious man that he was, Weatherwax felt pessimistic about ever really solving the problem. Earlier, Candolle had lamented, "I dare not hope that maize will be found wild." Weatherwax agreed. He saw no way of continuing the search, except by more study of the forms of the plants at hand.

Candolle had hoped that archeologists might dig up evidence of ancient wild corn. Weatherwax doubted it. He had seen many specimens of corn from old Indian campsites and burial places. In all important ways, it looked like modern corn, and that didn't seem to him very promising.

Paul Weatherwax left the door open for more details to be learned through the same kind of research he had been doing, but, so far as the main outlines of corn's past were concerned, he pretty much considered the case closed.

At just about the time that Weatherwax was writing *The Story of the Maize Plant,* another, younger man from the midwest was finishing his college study of plants and setting out in the world. The young man couldn't have known it, but he was beginning a career which would lead him to new theories about corn's history and to new discoveries which Weatherwax doubted anyone would ever make.

Just collecting evidence

Paul Mangelsdorf came to the Agricultural Experiment Station at New Haven, Connecticut, fresh with a bachelor's degree in agronomy from Kansas State College. The year was 1921, and a new chapter in the history of corn was just being written.

For more than fifteen years agricultural scientists and geneticists had been growing corn in the fields at the Experiment Station to investigate an odd phenomenon called *hybrid vigor*. Long before, Charles Darwin had noticed this phenomenon when he was experimenting with certain methods of plant breeding. If, for instance, he took the seeds from an ear of a good variety of corn and made sure that the plants which grew from those seeds were fertilized only by their own pollen, he was doing what is called *inbreeding*. The results of inbreeding can be very sorry, indeed. Immediately the offspring of the inbred plants are less productive than their parents. After several generations of inbreeding, the plants have greatly deteriorated. The stalks are shorter; the ears are smaller and fewer.

But if two different strains of wizened inbred corn are then cross-bred—that is, if one is fertilized with pollen from the other—the plants which grow from those hybrid seeds bounce back with tremendous vigor. They grow tall and lush and fruitful, and the ears may be bigger than those of either original ancestor.

How and why certain inbred and hybrid plants performed as they did was a problem on which several botanists, shortly after 1900, set to work with their new understanding of heredity. Much of the work was done at the Connecticut Experiment Station. At first it was chiefly a problem of genetics. Men wanted to figure out the scientific principles involved in hybrid vigor. But, after a while, they turned their attention to a practical question: How could scientifically controlled breeding be used to produce on farms the huge ears of corn that were grown in the experimental fields?

When Paul Mangelsdorf reached the Connecticut Station, botanists were on the point of making hybrid corn available to New England farmers, and Mangelsdorf joined in the continuing research.

Mangelsdorf's experimenting at the Connecticut Station was not the first work he had done with plants. His father was a seedsman in Atchison, Kansas, and the Mangelsdorf children (there were fourteen brothers and sisters) helped out. When he was six years old, Paul had a job pasting stamps on the seed catalogues his father mailed to customers. As he stamped, he began to wonder about the small, hard, unimpressive seeds themselves. It didn't seem possible that they could produce the multitude of bright-colored flowers he saw pictured in the catalogues. He said so, and his father gave him a batch of seeds to plant. Some of them grew just as the pictures promised. Others seemed to indicate that the catalogue artist had used his imagination. At any rate, the young experimenter had his first taste of the satisfaction that came from investigating something for himself.

When Mangelsdorf went to college, he got a job as assistant to a professor who was experimenting with grain improvement. There was far too much for one man to do, so Mangelsdorf took over the corn experiments and began the serious study of the plant which became his passion and his career. By the time he graduated he was well prepared for work in the hybrid corn project in Connecticut.

His experiments continued, and so did his studies. During the winter, when there was not much to do

at the Station, he went to Harvard University. As a student, Mangelsdorf could benefit from all the knowledge about corn and the nature of plants which scientists had been gathering. The theory of evolution and the main principles of genetics had been worked out. There was detailed information about the structure and workings of the maize plant and its relatives which botanists like Paul Weatherwax had pieced together. And a botanist could take his pick among a number of theories about the origin of corn.

At Harvard Mangelsdorf also found professors to stimulate his curiosity. One of them was a warm, explosive man named Edward Murray East who had started the hybrid corn experiments at the Connecticut Station years earlier. East encouraged Mangelsdorf to probe into the newest ideas of science. A push of a different sort came from W.J.V. Osterhout, a professor who remained skeptical about genetics and who poked fun at the young enthusiast for accepting so readily the modern notions of heredity. Half annoyed and half amused, Mangelsdorf decided to show the skeptic a thing or two about the importance of genes.

For six years Mangelsdorf commuted between study at Harvard and work at the Station. Near the end of that time he heard about an experimental attempt to cross corn with Tripsacum. He had given Tripsacum very little thought, but now he hunted up a specimen and examined it. Surely it was very different from corn, although the two plants might be more closely related than he had suspected. Here at the Station he had seen marvellous and fruitful results come from experiments with hybrids, and he decided to try his hand at hybridizing corn and Tripsacum.

But before this experiment could get well started, Paul Mangelsdorf—now Dr. Paul Mangelsdorf—was offered a new job. In 1927 he left Connecticut to supervise corn research at the Agricultural Experiment Station in Texas. New hybrids had to be developed from Texas corns, and the task was not easy. For a while he was kept so busy that the idea of a corn-Tripsacum experiment slipped from his mind.

The Texas Station had a number of branches across the state, and Mangelsdorf traveled periodically from one to another by train. One day he noticed some clumps of tall grass growing along the railroad track—clumps of Tripsacum. Again he thought of trying to cross it with corn.

By now he had an eager partner. Robert Reeves had just come to teach biology near by at Texas Agricultural and Mechanical College. Reeves was a cytologist, a specialist in the working of cells, and he shared Mangelsdorf's delight in corn research. The two men planned the corn-Tripsacum experiment together.

In a special plot they set out a few corn plants, and, when the silks began to glisten at the tips of the ears, they dusted them with Tripsacum pollen. Then they covered the ears to keep all corn pollen out. Soon the silks began to wilt and droop. This meant that the Tripsacum pollen was, indeed, at work thrusting its tubes downward toward the cobs. But, when the time came to strip away the husks, the cobs were almost bare. Only a few kernels had developed after all.

What went wrong? Perhaps, Mangelsdorf suggested, the trouble lay in the corn silks, which are much longer than those of Tripsacum. The Tripsacum pollen may have produced only a few tubes long enough to burrow the whole length of the corn silks. Next time, Mangelsdorf and Reeves decided to operate on their experimental corn ears when the silks were ready to receive pollen. First they sliced away the husks and snipped off all but an inch from each silk. Then, after dusting the stubs with Tripsacum pollen, they re-covered the cobs with artificial husks made of crepe paper.

The strategy worked. This time a good many kernels developed from the cross. Mangelsdorf wanted to make sure that these seeds would have every chance of germinating and growing, so he devised a special incubator in which to plant them. And the seeds did grow into hybrid corn-Tripsacum plants.

What had Mangelsdorf and Reeves proved? They certainly could not have said that corn—modern corn—would cross naturally with Tripsacum. They

*Three stages
in the experimental
crossing of
corn and Tripsacum*

really didn't know what they had proved, except that corn and this wild relative are not so distantly related that crossing is wholly impossible. (Later they learned of experiments indicating that corn hybridized even more readily with other varieties of Tripsacum.)

And what did their experiment have to do with the search for the origin of corn? Mangelsdorf and Reeves didn't know that either. But they kept on working. They crossed Tripsacum with corn again and again. They crossed corn and teosinte. They crossed crosses with crosses. They studied varieties of each of the three relatives in all their details. As they went along, they probed into the genetic makeup of their plants, and they learned from the work of others who did so, for they wanted to know exactly what genes inside the cells of corn were responsible for certain individual traits in the plant.

One trait that had puzzled Darwin, De Vries, and Asa Gray also intrigued Mangelsdorf and Reeves— the pods that surrounded the kernels in some varieties of corn. In ordinary kinds of corn these wrappings, the glumes, are unobtrusive and short. They appear only as part of the papery chaff on the cob. By now botanists knew that there was a single gene which influenced the growth of glumes. But there could be variations in this gene. If it varied in one way, the long glumes of pod corn would develop. If it varied in another way, the glumes would be the short kind found in ordinary naked corn.

The big unanswered question was: What did

these variants mean? Suppose the long glumes were the result of a mutation—a change in the gene normally associated with short glumes. This might mean that pod corn was a freak and, as Paul Weatherwax thought, of no importance in the search for the missing wild ancestor. On the other hand, a mutation might have brought about the *short-glume* trait. In that case, as De Vries thought, the long glumes of pod corn might be a holdover from the past when all corn seeds grew with individual wrappings. In other words, long glumes could be a primitive trait which lingered hidden in the cells of some modern corn and which reappeared from time to time. Was this possibility worth pursuing? Mangelsdorf and Reeves couldn't be sure.

As they kept working, an interesting pattern did begin to grow out of their investigations of ordinary corn and its relatives. After repeated comparisons of the three plants, Mangelsdorf and Reeves became suspicious of teosinte. It did not seem to be a really independent plant with unique characteristics of its own. Some of its traits were like those of corn; others were like those of Tripsacum; but none of teosinte's traits was essentially different from those of its two relatives.

For instance, the seed spike of teosinte resembles the spike of Tripsacum, but it grows separate from the male tassel, inside a kind of ear formation—as does the cob of corn.

Even the insects at the Texas Station treated teosinte as if it were a halfway-between plant. They ate

A corn-Tripsacum hybrid

corn most often and Tripsacum least; their appetite for teosinte was middling.

How could this in-between position of teosinte be explained? How could Mangelsdorf and Reeves fit it into any large picture of corn's past? And what did all the data they had gathered add up to?

Corn breeders from other parts of the country often visited the Texas Station to exchange information and discuss their problems. One day a visitor asked Mangelsdorf what theory he had about the origin of corn. Mangelsdorf replied that he didn't have one; he was just collecting evidence.

Some pieces fall into place

Late one night Paul Mangelsdorf was eating a snack in the kitchen. He had come from his laboratory at the Texas Experiment Station, where materials from his corn investigation were spread out on a big table so that he could study them—thousands of pieces making a great puzzle.

Mangelsdorf chewed and pondered. What sense did all this material make?

Then, between two bites of a sandwich, everything fell into place; suddenly the collector of evidence became a theorist. He had a new idea about the origin and evolution of corn and its relatives. Although the idea contradicted many of the things that other experts had been saying, he believed that the facts in front of him forced him to it.

There were a number of interlinking parts to Mangelsdorf's new theory. The first part explained the facts he and Reeves had gathered about teosinte. That plant seemed to be a combination of corn and Tripsacum—because it actually was. It had not evolved independently from some common ancestor of all three plants, as Paul Weatherwax insisted. *Teosinte originated as a hybrid—a cross between corn and Tripsacum.*

Two forms of pod corn
top: in the tassel
bottom: on the cob

Mangelsdorf did not ignore the difficulties he and Reeves had run into when they crossed modern corn and Tripsacum, but he speculated that a more primitive variety of corn might have crossed naturally with its relative.

If teosinte was a *descendant* of corn, it could not possibly have been corn's wild ancestor. Nor could Tripsacum have been the plant from which corn evolved. (On this point scientists generally agreed.) Therefore, *the ancestor of corn must have been corn.*

But what kind of corn? Unlike Weatherwax and many others, Mangelsdorf had not scratched pod corn off his list of clues to the identity of wild corn. Of course the doubly covered seeds in an ear of pod corn could not sow themselves any more easily than could ordinary naked corn seeds. But there was another variety that bore its glume-covered kernels at the base of the tassel. These tassel seeds had no thick jacket of outer husks to confine them. A tassel branch itself was fragile—it broke almost as easily as the stem of a wild plant that could sow itself.

Certainly modern pod corn was not identical with wild corn, but Mangelsdorf believed that its long glumes were a holdover from the past. Pod corn appeared now and then because of a rare primitive gene that survived in the hereditary materials of some present-day corn plants.

Although the glume-wrapped seeds of a wild maize plant may have been few and small, they did have value as food. Wandering people must have found nourishment in them. After men began to settle down and cultivate various plants, they undoubtedly brought wild corn into their garden patches. And at some point, said Mangelsdorf, those early farmers must have discovered a new form of maize with naked kernels growing among the old. A mutation—a permanent change in a gene—must have brought about a reduction in the size of the glumes.

In the eyes of farmers the new naked seeds would have been a pleasant development. The kernels were easier to get at; more important, the plant became more fruitful. The energy which once went into producing the pods now went into making bigger kernels. It was reasonable to suppose that even an unsophisticated farmer might have learned to choose the naked seeds for planting, in the hope he would get more of the same kind at the next harvest.

Of course the old kind of plant would not have disappeared all at once. It must have crossed and recrossed with the new short-glume corn, and the ancestral long-glume trait would have been passed along generation after generation. Although it became more and more rare, pod corn still would have continued to turn up now and then in farmers' fields.

Mangelsdorf wondered how these later people would have reacted when they came across one of the unusual pod-corn plants. Probably the best farmers would have considered the covered seeds a nuisance and would have weeded the pod corn out. Others might have saved and planted the strange seeds for good luck. Mangelsdorf thought there was evidence for this line of reasoning. He found that, in the part of Peru where ancient Indian corn farming had been most highly developed, pod corn never seemed to appear at all. In other areas, where tribes actually valued pod corn as a magical plant, it was common.

Mangelsdorf knew that modern pod corn is indeed freakish. Both the shaggy-eared kind and the kind that appears as tassel seeds are, in a sense, what botanists call monstrous. Pod corn is not a stable plant, and it cannot reproduce well or consistently. But, he said, these characteristics are to be expected in a plant that man has cultivated and crossed and carried from place to place for centuries. When the primitive pod-corn gene asserts itself in a modern corn plant, it is as though an old Model T Ford body were fitted around a powerful new V-8 engine. The contraption wouldn't get very far.

Both Mangelsdorf and Weatherwax believed that wild corn must have had its seeds protected by glumes, but they disagreed about pod corn. Whereas Mangelsdorf had come to think of the long glumes as a primitive trait, Weatherwax said they were an occasional modern development. Mangelsdorf's theory had one great advantage: It opened new leads in the search. Even if experimenters weren't sure, they could try out the idea that long glumes

might be a clue to the nature of wild corn still linger-
ing in the genetic makeup of the modern plant.
Mangelsdorf was confident enough of his facts to
state that *wild corn was some kind of pod corn.*

It was also *some kind of popcorn,* he said. Several
characteristics of modern popcorn seemed like traits
that wild corn might very well have had. The seeds
were hard and flinty, and usually quite small. The
particular toughness of the kernel—which has a
great deal to do with its ability to pop—would have
been a protection against insects and decay when
the plant had to survive in the wild.

Popping might also have been the very thing that
first led men to value corn as a food. Plucked ripe
from the plant, the hard little seeds wouldn't have
been very easy to eat. But perhaps someone happened to toss a dry stalk on a fire. He must have
been startled first by a series of sharp explosions,
then by the sight of the kernels puffed up into light,
appetizing morsels. Hungry men would have been
quick to repeat the experiment.

An eighteenth-century Spanish official named
Azara had reported that Indians in Paraguay did
prepare corn in almost that fashion. They took a
stalk of pod corn—the kind that has seeds in the
tassel—and dunked the tip in hot oil. It came out
with the kernels popped and still attached to the
tassel, looking like a spray of flowers that a lady
might wear in her hair. The popped kernels, said
Azara, tasted very good.

All these facts seemed to Mangelsdorf to point in

*Tassel-seed pod corn
dipped in hot oil
and popped*

one direction: *The ancestor of modern corn was some kind of wild pod-popcorn.*

Other observations that Mangelsdorf and Reeves had made during their experiments fitted together into another part of the new theory. In the process of corn's evolution, a rather thin and small seed spike had developed into the thick, tough, modern cob—a change no one had fully explained. Mangelsdorf now had an idea about what had happened.

He had spread out in the laboratory all kinds of cobs. Some were very thin and flexible, others so stout and rigid that they would break in half before they would bend much. He also had cobs from plants which were a cross between corn and teosinte, and these were generally tougher than ordinary corncobs. The chaff on the corncobs—that is, the remnant of glumes—was papery and flexible, but the chaff on the hybrid corn-teosinte cobs was stiffer and grating to the touch. And the more teosinte influence there was in the plants from which the cobs came, the tougher the cobs and the stiffer the chaff.

These facts could only be explained in one way. *Modern corn must be the descendant of a cross between primitive corn and teosinte.* What must have happened, said Mangelsdorf, was something like this: Somewhere, sometime, primitive corn crossed with Tripsacum to produce teosinte, which began to grow as a weed. Next, corn crossed with this new relative, teosinte, which possessed many characteristics of Tripsacum. Some of those characteristics were passed along to the new corn-teosinte hybrids. Finally, those hybrids in turn crossed again and again with corn, and as a result some Tripsacum characteristics became a permanent part of corn's genetic building material.

The main result of the inclusion of Tripsacum genes was a strengthened cob. This Tripsacum influence, said Mangelsdorf, did for corn what steel girders do for a skyscraper—it provided a strong frame, which allowed the ear to grow larger and more rigid, and to be able to bear more kernels.

So Mangelsdorf now had what he and Reeves called a tripartite (three-part) theory about the origin of corn:

1   The ancestor of the modern corn plant was a pod-pop variety of corn (that is, it had glume-covered kernels which would explode when they were heated).
2   Teosinte is the result of a natural cross between primitive corn and Tripsacum.
3   Modern corn got some of its important characteristics from Tripsacum genes. (These came to it through crossing with teosinte.)

Many questions still remained unanswered. In what part of the world did wild corn grow? Where and when did corn cross with Tripsacum? How and where did men learn to grow corn? And what happened to wild corn after men became farmers?

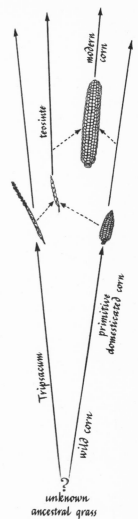

The family tree of modern corn, constructed by Mangelsdorf and Reeves (1939)

The where of it all

To make the story complete, Paul Mangelsdorf wanted to know *where* the maize plant first grew wild. Many botanists—perhaps most—agreed that it was a native of the New World. Yet no one had real proof of its geographical origin, and some investigators believed there was evidence that it grew in various places outside the Americas before Columbus.

Part of the confusion started with Columbus himself. Members of his first expedition certainly found cornfields on one or more of the Caribbean islands, and they reported that the Indians called the plant something that sounded like "ma-hees." (In Spanish it was spelled *maiz,* and that is where we got the word *maize.*) But in all that remains of his writings, Columbus failed ever to state clearly that he brought corn back with him to Spain. One passage in a letter has even been interpreted to mean that he had seen the plant growing in the Spanish province of Castile *before* he sailed. To add further complications, he insisted that his voyages had taken him to Asia, and so everything he discovered was "Asian." After his time, the idea lingered in Europe that corn was really an Old World plant.

The first European scholar to include a picture of corn in a book about plants—in 1542—labeled it *Turcicum frumentum,* which is Latin for "Turkish grain." At that time trade with the Far East had brought many curiosities to Europe, often by way of Constantinople, in Turkey, and it was common to say that anything odd, foreign, and new was "Turkish." (One of the strangest looking birds in creation came to be called a turkey.) Whatever its origin, the name "Turkish grain" or "Turkey wheat" was used for a long time in Germany, Holland, Italy, and England. In some parts of Spain that is what the people still call it.

But if corn was Turkish, the Turks did not claim it. They called it "Egyptian grain." The Egyptians called it "Syrian grain." To the Greeks it was "Arab grain"; and on the island of Sardinia it is even now called "Moorish wheat."

Under one name or another, corn was growing in many parts of the Old World by the time sixteenth-century geographers had learned where to put the new continents on their maps. Within twenty-five years of Columbus' first voyage, Portuguese explorers had taken it to India and the Spice Islands. Within seventy-five years a picture of corn appeared in a Chinese book of natural history.

On the other hand, it was quite possible for educated Europeans to be completely ignorant of the plant more than a hundred years after Columbus. When the Pilgrims first landed on Cape Cod, they

TVRCICVM
FRVMENTVM
Turckiſch korn.

*"Turkish Grain"—The first picture of corn to appear in a European
botany book*

dug up a cache of Indian baskets "filled with corne; some in eares, faire and good, of diverce collours, which seemed to them a very goodly sight (haveing never seen any shuch before)." So wrote William Bradford, one of the *Mayflower's* passengers.

In Britain *corn* was a word that meant simply *grain*—any kind of grain. It is still used in that way. If an Englishman talks about corn he means the major grain crop of any particular area. In England a cornfield is a field of wheat; in Scotland it is one of oats. In the American colonies the main cereal food was the seed of the maize plant, which early settlers called "Indian corn" at first, later shortening the name to "corn."

By the nineteenth century, when botanists began to pay more attention to the geographical origin of plants, several facts seemed obvious. The maize plant (called by scientists *Zea mays*) grew almost everywhere in the New World between the Great Lakes and southcentral Chile. It was the major grain crop of both the Indians and the European settlers, but certainly the Indians had been skilful corn farmers long before Columbus arrived. So far the evidence pointed straight to an American origin.

Then, in 1810, a civic-spirited Italian named Molinari put a new twist into the record. That year he published a history of the town of Incisa, where his family had lived for generations, and he quoted documents that seemed to show maize in Italy two hundred years before Columbus was born! Among

Incisa's notable citizens, said the author, were an ancestor of his and another man who had gone on a Crusade against the Moslems at the beginning of the thirteenth century. They had fought in a great battle near Constantinople, then returned to Incisa in August, 1204, with gifts for their fellow townsmen. Their trophies included a piece of the true cross and a bag of white and yellow seeds identified as corn.

This story seemed all the more convincing because in parts of Italy corn was indeed called "Turkish grain." For what it was worth, the French botanist Mathieu de Bonafous put a discussion of the Turkish seeds into a book about corn which appeared in 1836. He also reported that an explorer who opened up an ancient Egyptian tomb near Thebes had discovered an ear of corn among relics thousands of years old.

Some time after Bonafous' book was published, a professional French historian took up the discussion. He didn't care much about corn, but he liked facts, and he had his doubts about the tale of the Crusaders from Incisa. True, Italians on this particular Crusade had attacked Constantinople. But they had not even entered the Turkish province, where supposedly they found corn, until *after* August, 1204. Further checking disclosed more wrong dates. There were other things that could not be checked because Molanari had said he was quoting from records which had once existed but which scholars knew had now been lost.

In the end the history of Incisa proved to be largely a fraud. Probably its author had amused himself by inventing a distinguished ancestor; and, while he was at it, he had made his home town seem to be the first in Europe to have seeds of corn.

And so the "Turkish grain" story could be eliminated when Alphonse de Candolle came to do his careful work on the origin of cultivated plants. He could also dismiss the report about corn in Egypt before Columbus. No recent excavators had dug up more of it. A search of Egyptian art revealed pictures of many grains, vegetables, and flowers which had been grown long ago—but nothing that looked like corn. That lone specimen in Egypt could only be one thing: an ear of modern corn which some joker had slipped into the ancient tomb for the French explorer to discover.

Scholars found no Sanskrit or Hebrew words for corn and no description of it in the Bible. The word *corn* does appear in the King James version of the Bible, but that is because the seventeenth-century translator used the common English synonym for *grain*.

Paul Mangelsdorf and Robert Reeves sifted through many of those old investigations while they prepared a book about their own researches and the tripartite theory. Their title, *The Origin of Indian Corn and Its Relatives,* suggested that they believed corn was a native New World plant domesticated by Indian farmers in the Americas before Columbus

Earliest known Chinese picture of corn. The caption reads "gemlike grain from western China."

arrived. With that assumption they turned to a question which was much more difficult to answer: What part of the New World was the original home of wild corn?

Botanists then had two different approaches to the problem of the geographical origin of plants. Some favored the idea that a plant must have originated in the place where it could be found growing along with the greatest number of its relatives. For corn, that meant Central America and southern Mexico, which were centers of ancient Indian farming, and which were also the areas where corn existed along with teosinte and Tripsacum. If all three plants evolved separately from a common ancestor, as Weatherwax believed, this region might well be called the first home of maize.

Other botanists thought that a plant originated in the place where it had developed into the greatest number of different types. This theory would direct a corn investigator to South America—to Peru, where pre-Columbian agriculture reached its highest development. There corn is found today in magnificent variety.

Which area was the right one?

Mangelsdorf and Reeves said South America. In addition to subscribing to the variety theory, they had further reasons for their choice. Because they believed that teosinte is a cross between corn and Tripsacum, they had no need to look for corn's original home in the place where teosinte now grows.

They argued that corn had been carried by the Indians from South America to Middle America (Mexico and Central America), where it crossed with Tripsacum, which grows abundantly there.

Because Mangelsdorf and Reeves thought that the long glumes of pod corn were a genuine ancestral characteristic, they might have expected to find it often in fields near the place of origin. It did turn up in the lowlands on the eastern side of the Andes Mountains, but on the western side of the Andes, in the highlands, where Peruvian Indians were most accomplished farmers and grew many different varieties, pod corn was missing.

Mangelsdorf thought it reasonable to suppose these highland farmers had successfully weeded pod corn out. Still, he wanted evidence that pod corn had grown there. He knew that Indian artists often decorated their pottery with realistic models of corn, so he began searching through collections of Peruvian artifacts. And one day in the Peabody Museum at Yale University he found exactly what he was looking for: a piece of clay sculpture from Peru which looked to him like nothing so much as an ear of pod corn.

In 1939 Mangelsdorf and Reeves published *The Origin of Indian Corn and Its Relatives;* and, as they later observed, the book stirred up a great deal of valuable discussion. But it also opened what they called "a veritable Pandora's box of unrestrained speculation" about one point they had thought

Ancient Peruvian clay model which Mangelsdorf believed represented an ear of pod corn

safely settled: Was corn really a New World plant after all?

Some of the speculation was based on botanical reports. Plant collectors, explorers, and travelers had found maize in unexpected places. They had come upon it in the hill country of Burma, in the mountains of Sikkim, in the interiors of the islands of Formosa and New Guinea. In all these places corn was grown by rather backward inland peoples —not by the more sophisticated farmers near the seacoasts. This seemed very curious. If such a valuable crop as corn had been brought to Asia by Europeans, why hadn't it taken hold along the coasts where Europeans had first landed? Why had it skipped across into the hinterlands?

At one remote spot in the mountains of Java, a group of people raised corn, although their neighbors did not. Near this spot were ancient ruined buildings that resembled certain prehistoric temples in Mexico. Moreover, among these people were dwarfs who had a striking appearance. They resembled a kind of dwarf often carved in stone by certain pre-Columbian Mexican sculptors. It seemed that someone, sometime, must have carried corn seeds either from America to Asia or from Asia to America.

The experts on American Indian culture also took sides in a vigorous argument. Those on one side believed that there had been no important contacts between the Old World and the New World before Columbus. This meant that the Indians had developed their own civilizations, their own food plants and agricultural techniques, without help from men of any other continents.

There were, however, hints that plants had spread outward from the Americas before the time of Columbus. Some ancient Nigerian pottery looked as if its maker had decorated it by rolling a corn-cob over the wet clay before firing. Sweet pota-toes seemed to have traveled from South America to islands far across the Pacific Ocean. In 1947 six young men sailing on the raft *Kon-Tiki* proved that Indians could certainly have floated westward from Peru to reach the islands. Did some ancient raft carry men and corn across the Pacific?

Scholars also took an interest in the voyages that Arabs made before Columbus. It began to appear that Moslem science might have given sailors the tools and skills they needed for navigating the South Atlantic long before 1492. Was it possible that Arabs had trafficked in a variety of corn which grew here and there in the Moslem world in pre-Columbian times? If so, then there might be an easy explanation for the names "Turkish grain" (Turkey was a Moslem center) and "Moorish wheat."

Other experts maintained that they had evidence of traffic from Asia to the Americas across the Pacific Ocean long before Columbus' day. For example they pointed to many carvings in Mexico that closely resembled carvings in Asia. The Aztecs in

Where corn was grown in the New World at the time of Columbus

Mexico had played a game almost exactly like the game called parcheesi, which was played in Asia long before 1492. Mirrors in Central America seemed to resemble mirrors in China. In both areas craftsmen had made a very special type of three-legged vessel with a bird decoration on its conical lid.

It was not impossible for men to sail across the Pacific in ancient times. After all, the art of navigation did not really start with Columbus and the Portuguese. Long ago the Polynesians learned to guide themselves by the stars. They could and did steer a course to Hawaii when they chose. A route to the coasts of the Americas might have been no more difficult a problem.

Could men have brought corn seeds to the New World from Asia? There were some botanists who still said it was possible, in spite of all the arguments against it which men like Mangelsdorf and Reeves offered. Paul Weatherwax was among those who staunchly agreed that corn originated in the New World. On many other questions he just as staunchly disagreed with Mangelsdorf and Reeves. He was skeptical of the whole tripartite theory, and he said so.

Theory and practice

The trouble with Mangelsdorf, said Paul Weatherwax, was that he had too much theory and not
enough fact. His whole idea was like a rickety
house. If one part could not stand up, the entire
structure would collapse.

Not so, replied Mangelsdorf. The tripartite theory
did fit together neatly, but it didn't have to. The
parts did not depend upon each other. Any one of
them could still have been true if the others were
proved mistaken. Wild corn could have been a kind
of pod-popcorn even if teosinte were not the descendant of a corn-Tripsacum hybrid; and, no matter where teosinte came from, it still could have
crossed with corn.

Weatherwax came to agree that some modern
corn had, in fact, inherited traits from a cross with
teosinte. But he argued adamantly against the
theory of a cross between corn and Tripsacum. He
pointed out the difficulty Mangelsdorf and Reeves
had had in making the cross, and he insisted that
such a farfetched notion was just not necessary to
explain the facts.

Mangelsdorf repeated his statement that he wasn't
talking about a cross between *modern* corn and its

relative. Then he applied Weatherwax's own rule by pointing out that the tripartite theory accounted for more facts than did the Weatherwax theory about the independent evolution of teosinte, which could not explain why teosinte seemed to have no real traits of its own. The whole question of teosinte's in-between position, Weatherwax answered, wasn't really very important.

Again Weatherwax gave reasons why he thought pod corn should not be called "wild corn." Mangelsdorf repeated *his* argument: He hadn't said that modern pod corn was the wild ancestor—only that wild corn was *some kind of* pod corn.

It was still Weatherwax's opinion that Mexico and Central America must have been the original home of corn. Mangelsdorf conceded that there was some evidence of a Middle American origin, and he abandoned his earlier belief that the lowlands of South America had been the *only* center from which cultivated maize had spread outward. But he argued that there might have been more than one center of origin.

On one point, however, everyone agreed: More facts had to be found before any botanist could give the final answer about the origin of corn. The problem was still *where* to look for it. Mangelsdorf shared the pessimism of Weatherwax—and of Candolle before him—about the possibility of finding ancestral wild corn growing. Weatherwax doubted that even the remains of the missing ancestor would ever be discovered. But Mangelsdorf agreed with Candolle that archeology might offer some hope.

Mangelsdorf was especially interested in ancient garbage. He knew that there were many sheltered trash heaps in front of caves in Mexico and the southwestern United States which had once been inhabited by Indians. In the dry climate those places remained almost as they had been left, many hundreds of years before. Archeologists had often dug down through such refuse and had found undecayed parts of plants, animals, even human bodies, along with lost or broken tools, jewelry, weapons, and household goods.

Unfortunately most archeologists paid little attention to ancient plant material. Their expeditions were usually financed by museums eager for dramatic and interesting man-made articles to put into collections and display cases. A dried-up corncob isn't much to look at. So, when crews finished investigating caves, they often burned the uninteresting plant rubbish or threw it out where the next storm would wash it away.

On one expedition the diggers reported that their mules ate nine-hundred-year-old corn from an Indian trash pile. Another report told of some small, unusual kernels that an archeologist had found in a cave in New Mexico. "Each grain," he said, "seems to have a little horny envelope of its own." Was this ancient pod corn? Might it even have been the remains of wild pod corn? Perhaps, but no one would ever know, because the specimens had disappeared.

Archeology is a science in which things have to move rather slowly. The careful work of the excavator takes time, and he tends to become a specialist, bent on solving one mystery at a time. It simply did not occur to most people that domesticated corn itself is man-made and well worth study as an artifact.

So archeology and corn science kept to their separate ways for a number of years after Mangelsdorf and Reeves published their book. Mangelsdorf went back to Harvard University as professor of botany; then he became director of the Botanical Museum there. In 1948, he made a start toward bringing the two sciences together. That year a young student of archeology, Herbert Dick, set out with an expedition from Harvard to excavate Bat Cave, in New Mexico. And Mangelsdorf saw to it that a botanist went along.

Evidence in Bat Cave

Bat Cave is a great gouge in a cliff 165 feet above the Plains of San Augustin in New Mexico. Chunks of rock have fallen from the cavern's high roof, and the floor is covered with wind-blown dust. Before Herbert Dick and his small party arrived there, treasure hunters had already dug into the cave floor, searching for old Indian pottery or ancient turquoise jewelry.

Dick and the men and women in his team went to work in some undisturbed side passages and in the rubbish heap at the front. During two summers they shoveled dirt and sifted it through wire screens. Often they had to work more carefully, using trowels and whisk brooms to avoid breaking some fragile object. No matter how cautious they were, clouds of dust filled the cave, so they had to wear masks with air filters that looked like pig snouts.

The work in Bat Cave was uncomfortable, but it brought results. The diggers turned up pieces of pottery, remains of ancient meals and campfires—and corncobs. Not only cobs, but also tassels, leaves, husks, and kernels.

C. Earle Smith, the botanist on the expedition, packaged and labeled these specimens for Mangelsdorf to study in his laboratory at Harvard.

Down at the bottom of a rubbish heap the diggers finally came across some bits of dried plant material which they could not identify. The stuff hardly seemed worth saving, but Smith kept it and sent it along anyway.

When the packages arrived, Paul Mangelsdorf was delighted. And the small items which had stumped the archeologists seemed almost too good to believe. Mangelsdorf's trained eyes told him what they were—tiny primitive corncobs. One of them even had part of a single hard kernel still in place. Now at last collaboration between botanists and archeologists had produced something new in the long effort to solve the mystery of corn.

Of course Mangelsdorf wanted to find out all there was to know about the dry wizened cobs. What had they looked like when the corn was first picked so long ago? His friend and associate Walton Galinat, a specialist on the structure of corn, joined him in the investigation. For an entire week the two men worked together dissecting a cob, measuring each part, and checking one another at every step.

The smallest cob found in Bat Cave

The remaining bit of a kernel was very hard— like popcorn. And some popped kernels had indeed been among the Bat Cave trash. Around the places on the tiny cob where the kernels had grown were the remnants of what must originally have been long glumes. This primitive corn was pod-popcorn.

Finally Galinat was able to draw a diagram of the

cob, and it showed what Mangelsdorf had so often seen in his mind's eye: Tiny hard kernels of popcorn, each with a wrapping of rather long glumes, were attached so lightly that they might almost have been able to break off and sow themselves after ripening.

` Mangelsdorf also examined husks that came from deep down in the cave's rubbish heap. They did not look as if they had fitted very closely around the small cobs. Perhaps husks of this kind gave some protection to the corn while it was growing. Then, later in the season, they may have drooped and spread apart, freeing the ripe ear.

Was this wild corn at last?

Mangelsdorf and Galinat and Smith thought not. Primitive though it looked, it probably could not have grown without man's help. More likely the Bat Cave people had got both their knowledge of farming and their corn seeds from other people who lived farther south. The botanists preferred to call the small Bat Cave corn a primitive form of domesticated corn—the nearest thing to a wild plant yet discovered.

Herbert Dick's expedition turned up many kinds of corncobs in addition to the tiny ones. Mangelsdorf was interested in them all. The archeologists had dug down very carefully through one layer of rubbish after another, and they knew that the cobs they found in the top layer were the most recent. As they went deeper and deeper in the rubbish, they

An artist's reconstruction of Bat Cave corn (actual size)

Diagram of the structure of Bat Cave corn (actual size)

Cob and husks of
Bat Cave corn
(one-half actual size)

went farther and farther back in time, until, in the bottom layer, they reached the oldest cobs of all.

Every specimen was given a label that identified it. Now, when the digging was done, the scientists could arrange the Bat Cave cobs in order, according to the layer in which each one had been found. The smallest had come from the oldest layer, and the cobs at the next level were only a little larger.

Then, at higher levels, about halfway along in the history of the cave, much larger cobs suddenly began to appear. They were stronger, and they held more rows of kernels. The glumes around the kernels were smaller.

Some of these more recent cobs had sharp, tough ridges around the places where the kernels had been attached. Mangelsdorf remembered that similar stiff ridges appeared when he crossed corn with teosinte. Here, he felt sure, was actual historical evidence to support his belief that teosinte had played a part in the evolution of corn.

(Later, Mangelsdorf made a demonstration to emphasize this point. First he hunted among the cobs of experimental plants that came from corn he had crossed with teosinte. He found some that were almost exactly the same size and shape as ancient cave specimens and about equally rough. Then he disguised these modern cobs by boiling them in oil to color them a dark ancient-looking brown. When he compared his teosinte-influenced cobs with the really ancient ones, it was almost impossible to tell them apart.)

But teosinte did not seem to grow near the Plains of San Augustin. So the Bat Cave cobs must have got their toughness from some foreign source. Possibly, after farming had begun near the cave, traders had brought in new seeds from an area where teosinte was common. When this new variety crossed with the old Bat Cave corn, the teosinte influence began to show.

In the upper layers of the cave rubbish, still different types appeared. Some of them closely resembled modern corn.

So the whole collection of cobs gave step-by-step evidence of evolution in corn, from ancient to near-recent times. Mangelsdorf was pleased—especially because the evidence seemed to confirm two of the points in his theory about the origin of modern corn: (1) that wild corn was a pod-popcorn and (2) that some of modern corn's important traits came from teosinte.

The archeological record showed *how* corn had developed in the Bat Cave region. The next thing to find out was *when*. In the top layer of rubbish the cobs were mixed with broken bits of pottery of a kind which, archeologists knew, Indians in that part of New Mexico had made between the years 500 and 1000 A.D. So the corn in the top layer must have been growing near Bat Cave at about that same time.

Although there were no potsherds in the bottom layer, the scientists did find other clues. The lowest layer rested on a bed of dust and sand which a geol-

Spear points, yucca leaf basket, braided animal-hair cord from Bat Cave

ogist said had blown into the cave sometime before 2500 B.C.

Still Mangelsdorf and Herbert Dick wanted more accurate dating. Fortunately they didn't have to wait very long for it. By 1949, Willard Libby, the atomic scientist, had developed carbon-14 dating. This process, which measures the radioactive carbon in such things as wood or bones or corncobs, can tell with fair accuracy the length of time a plant or animal has been dead. Best results come from material that has been charred in a fire.

Herbert Dick sent to a carbon-14 laboratory some charcoal he found associated with the tiny cobs in the lowest layer of Bat Cave rubbish. The laboratory gave him a date that was even earlier than the geologist's estimate. If the people who had built the fire in the cave had indeed tossed the cobs onto the cave floor, then Indians were raising corn in New Mexico about 5600 years ago. That was impressive. But it still didn't tell when (or where) corn had grown wild.

While Mangelsdorf, Galinat, and Smith were putting together their story of corn in Bat Cave, another archeologist, Richard MacNeish, took up the search for the missing ancestor. MacNeish was probing into caves farther south—and trying to dig farther into the past.

Devil's Canyon corn

Through lonely mountain country in northeastern Mexico runs Cañon Diablo—Devil's Canyon. Scrubby desert plants grow along its slopes. At one spot a side canyon branches off.

It was here, people told Richard MacNeish, he should begin looking for a cave in the rocky wall. In March, 1949, he searched, and a little way up the side canyon he finally found it. La Perra Cave was not the perfect place for an archeologist to dig—high in the cliff wall, hard to reach, fifteen miles from the nearest drink of water. Nevertheless, MacNeish decided to give it a try. Perhaps Indians had lived in this unlikely place long ago at a time when Devil's Canyon had a better climate.

The archeologist and a crew of helpers went to work. They did find evidence that people once lived in the cave, but those early settlers apparently hadn't left any interesting trash. MacNeish soon gave up. He had other things to do, so he asked the foreman of his digging crew to pack up the equipment they had been using at the cave and meet him two days later.

The foreman came to the rendezvous on schedule. But he confessed to MacNeish that he hadn't packed up the gear. "Why the devil not?" MacNeish demanded. The foreman grinned and handed over something he had discovered after MacNeish left: three small, ancient corncobs.

The digging began again. Every day MacNeish's party made the dangerous trip in and out of La Perra Cave, inching their way for more than a hundred yards along a high narrow ledge. They shoveled and sorted material and turned up more old corncobs, some of which were very small and very ancient indeed. Each specimen was put into a labeled bag that would go to Paul Mangelsdorf and Walton Galinat at Harvard.

At last the party finished exploring the cave they had almost passed up. Just before their final dizzy trip out across the ledge, they made still another discovery: There was a much easier, safer way into the cave from the top of the cliff!

After MacNeish's corncobs were arranged in order, they made a clear picture of evolution, much like the picture that had come from the Bat Cave cobs. La Perra Cave corn was perhaps not so old, but it told a fascinating story. When the scientists began to study it, they could see not only how corn had developed but also how people's use of it had evolved.

A few of the small early specimens had been mashed until one could hardly tell what they were. They looked exactly as if they had been chewed—husks, cobs, and all. The scientists decided to experiment. They picked some fresh, very young corn and chewed it without any cooking. It was, they reported, "tender, succulent and sweet." Small ears of raw, juicy young corn could have been an important food before people learned how to cook it.

Some of the cobs from La Perra Cave were slightly burnt. This might have happened after people discovered that dry corn would pop. All they had to do was sharpen a stick, poke it into the end of the cob, and hold the ear over a campfire. Again the scientists experimented. They held an ear of modern popcorn over an electric hot plate in the laboratory. The kernels burst open while they still clung to the cob. The smell was good, and the popped kernels came off easily. This test cob was burnt a little, too, like the prehistoric cobs from the cave. Mangelsdorf was reminded of his guess, made years before, that men might have thought corn worth cultivating once they discovered it would pop.

Several other cobs from the cave had little bits of the kernels still clinging to them—as if the ears had been roasted in the husk and then chewed, much as we roast and eat sweet corn at a picnic today.

Cobs from the upper layers of the cave looked battered, as though someone had rubbed them or beaten them to get the kernels off. This is what people would have done after they had learned to grind dry corn into meal. And grinding stones had, indeed, turned up in these top layers of cave debris.

La Perra Cave told Richard MacNeish much that

Chewed ears of corn found in La Perra Cave (actual size)

he wanted to know about the development of Indian culture. It had been a long road from ear chewing to corn grinding. The first farmers to use the cave had treated their crop almost as though they were still primitive wandering food gatherers who had not yet learned agriculture. They had plucked ears that were still growing and had eaten them raw. Later Indians had learned to wait until the crop was ripe, and then to store the dry corn, which could be kept all winter, ground into meal, and eaten when it was needed.

The cobs from La Perra cave also gave Mangelsdorf new material to work with. None of them seemed to be as old as the oldest Bat Cave specimen, but they helped fill out the picture of the development of corn under domestication. Mangelsdorf found that the cobs resembled in many ways three races of maize which are grown in Mexico today. And there was one special cob, not quite like anything he had seen before. Its glumes were long, though they probably didn't cover the seeds completely. It had eight rows of kernels in a very loose formation. Mangelsdorf, MacNeish, and Galinat studied this cob (number 127D5) carefully and came to the conclusion that it was even closer to a self-sowing maize than the oldest Bat Cave corn had been.

Corn-grinding tools from La Perra Cave

But it wasn't wild corn, and the search had to go on. After 1950 MacNeish spent as much time as he could roaming over the Mexican countryside, digging in the north and in the south. He crouched through so many caves that he said he was beginning to look like a Neanderthaler.

Meanwhile some other diggers were at work with modern machinery in the heart of Mexico City. They were not looking for corn or for fossils, but they made an unexpected discovery about both.

Fossils

Could fossil maize be found? Many a corn investigator had hoped so. An ear of wild corn or a whole plant that had been buried and turned to stone would show what wild corn had looked like and where it had grown.

When Darwin was a young man he thought he had discovered fossil ears on an island off the coast of South America. But they turned out to be nothing more than leftovers in a refuse pile at an old Indian campsite.

A more promising lead appeared in 1914 when a doctor from St. Louis found a peculiar ear in a curio shop in Peru. It was short and fat like the corn that grew in the nearby mountains, and it was rock hard —like a fossil.

The doctor bought the little object, which eventually found its way to the Smithsonian Institution in Washington, D.C. People there looked at it and said it was a fossil. They even gave it a scientific name— *Zea antiqua,* which means *ancient corn.*

This lone specimen of *Zea antiqua* sat in the museum for almost fifteen years. Then a corn expert examined it and had doubts about it. For one thing, it made a slight noise when he picked it up—as if it

76

had loose parts inside. The expert got permission to dissect it.

In the museum laboratory a man with a steady hand and a sharp saw went to work. He took one slice off the hard little ear, then another. The "fossil" was nothing but baked clay, and a hollow place inside it held three tiny clay balls. An Indian had modeled the ear long ago, perhaps as a rattle for a child, but more likely for use in a religious ceremony of some kind.

From time to time, reports of other petrified corn came to Paul Weatherwax at Indiana University. But when he examined the specimens, they turned out to be either clusters of mineral crystals or a special kind of rock formation called a *geode*.

The years went by, and nobody found an authentic fossil. Maybe corn was such a new plant on earth that there hadn't been time for any of it to become fossilized. Or perhaps it hadn't grown in the right spots. Plants in wet or swampy places are most likely to be turned into fossils, but botanists thought that wild corn must have grown best in somewhat high, dry country.

Then came the accidental find in a buried lake. Its story began one day in 1950 when a team of engineers from Switzerland went to work on a busy street in Mexico City. They were getting ready to put up a new building, and they wanted to know whether the earth was solid enough to support it. To gather this information, they used a drill that could

Zea antiqua, *the false fossil ear of corn*

*Rock formations
which have been
mistaken for fossil corn*

bore into the ground and bring up samples. As the traffic sped by, the drill went deeper and deeper, down more than two hundred feet under the city.

At that same time, a group of scientists from the United States happened to be working near Mexico City. Among them was the botanist Paul Sears. By chance Sears heard what the Swiss engineers were doing, and he rushed into the city to see them. He wanted the drill cores—the cylinders of earth that they were bringing up from underground.

Sears had to do a great deal of persuading, but he finally got the cores. He made careful notes on the depth from which each one came; then he took them to the United States, where they could be studied in a laboratory.

Sears was working in a new science called *palynology*, which is the study of pollen. He hoped that the cores from deep underground would contain fossil pollen.

Mexico City, Sears knew, is built where there was once a lake. Every year the flowering plants that lived near this lake shed billions of pollen grains. Some of the grains did their job of pollinating flowers. The rest were blown here and there until they settled on land or water. Those which fell in the lake slowly sank to the muddy bottom.

Streams washed sand and mud into the lake to cover the pollen. The next year more pollen fell, and more soil was deposited on top of it. This went on for thousands of years. The lake grew shallower and smaller. Eventually men drained the last of it and turned it into dry land. Buildings went up where there had once been water for pollen to fall into.

Pollen is strange stuff. Its outside layer is made of a tough substance something like plastic. In the mud at the bottom of a lake it can remain whole for tens of thousands of years, even after the mud hardens into rock.

Pollen's tough covering makes it possible for a palynologist to find the individual grains and study them. First he puts his sample of pollen-bearing rock into an acid bath. The acid eats away the minerals, but it does the pollen no harm at all. Next the scientist washes the pollen grains, dyes them, and puts them into a bed of clear jelly on a piece of glass. Now he can study them through a microscope.

Each plant's pollen is as identifiable as a man's fingerprints. An expert can look at a sample and tell what plant it came from. Fresh corn pollen is smooth and rounded, but when it dries it looks more like a football bladder with the air let out.

Fossil pollen did appear in the drill cores that Paul Sears brought home from Mexico City. The drill had gone down through layer after layer of the ancient lake bed, and of course the deeper it went the farther it reached back in time. About two hundred feet down there was pollen from plants that had bloomed near the lake eighty thousand years ago.

Some of the pollen in the deep layers had been

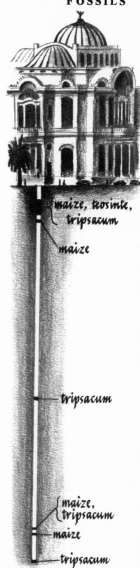

Levels of fossil pollen in the Mexico City drill cores

*Pollen grains
greatly magnified
top: oak
center: dandelion
bottom: rose*

large and smooth. It caught the eye of Kathryn Clisby, who was working with Sears. These grains, Mrs. Clisby was sure, must have come from a plant in the grass family. But which plant? The two scientists thought it might have been teosinte.

Did teosinte grow in Mexico eighty thousand years ago? If it did, then Paul Mangelsdorf would have to change his opinion that it had developed only after men began to cultivate corn. Naturally, Mangelsdorf was interested to hear about the drill cores; and he sent Mrs. Clisby some samples of modern teosinte pollen so that she could compare them with the ancient fossil grains.

The new and the old pollens did resemble each other, but not exactly. The measurements of the fossil grains were more like those of corn pollen. Was it possible that these specimens had come from a corn plant? Clisby and Sears could hardly believe it. Again and again they measured and examined. Finally they decided to consult another botanist who specialized in fossil plants. They sent the material to Harvard University, where Elso Barghoorn and his assistant Margaret Wolfe examined it. These two went over the evidence with great care and gave their opinion: The ancient pollen undoubtedly had come from a corn plant.

At last scientists had seen a scrap of true fossil corn. Moreover, this must be *wild* corn pollen, because there were no farmers—in fact there were no people at all—in the Americas eighty thousand years ago. And so an old and much discussed question about corn's origin had now been settled once and for all. Human beings could not have brought the original maize plant to the New World from Asia or Africa.

The experts compared their fossil pollen grains not only with modern corn pollen but also with ancient dried-up pollen from Bat Cave. The three types were so much alike that one could hardly tell which was which. No doubt about it—*corn had been corn in the Valley of Mexico for at least eighty thousand years.*

After the Bat Cave discovery, Paul Mangelsdorf had begun to suspect that wild corn might have grown in Mexico as well as in South America, where he had first thought it originated. The fossil pollen now indicated that the wild plant had indeed been there for Mexican farmers to find and domesticate. Still, Mangelsdorf said, men in more than one place might have learned independently how to be corn farmers. He would continue to believe that South America was one possible center of origin.

On the other hand, the drill cores helped to confirm Mangelsdorf's ideas about teosinte. In the samples that came from great depths, there was no teosinte pollen at all—only that of corn and Tripsacum. But, nearer the surface, teosinte pollen and abundant corn pollen appeared together—and in layers of mud that were probably deposited after Indians became farmers. This could mean that teo-

fossil corn pollen, 80,000 years old

dried pollen of modern corn

fresh pollen of modern corn

teosinte pollen

tripsacum pollen

sinte developed as a hybrid after corn was domesti-cated—just as Mangelsdorf had been thinking for a long time. Obviously teosinte could not have been corn's wild ancestor. And it seemed less likely than ever that teosinte had evolved, along with corn and Tripsacum, from a common ancestor.

Of course, the teosinte pollen might be explained in another way. The plant might have developed in-dependently in some other place, then started to grow in the Valley of Mexico only at the time when its pollen showed up in the drill cores. But that was something Mangelsdorf would believe only when he saw tangible proof. In the meantime, he had new experiments to do. He was at work trying to *create* corn which would grow wild.

Wild corn reconstructed again

There was a garden patch in Boston, a few miles from Harvard, where Paul Mangelsdorf grew corn every summer for years. He and Walton Galinat always had a number of experiments in progress in this little field. With one batch of plants they might try to discover the relationship between the size of an ear and the size and position of the leaves on the stalk. Another batch might help to determine whether the weight of tassels had anything to do with the weight of ears.

One special group of scrawny plants got very careful attention summer after summer. With these the botanists were trying to create something resembling wild corn. As a result of his work in Texas, Mangelsdorf believed that modern corn must still have some of the primitive genes which had determined the traits of its wild ancestor. The pod-corn characteristic was, he thought, evidence of one such holdover from the past. There must be others, too.

If an experimenter could identify the primitive traits which show up in modern varieties of corn, he could learn from them. Perhaps, by cross-breeding and selective breeding, he could put enough of these traits together in one plant to reconstruct wild corn. This would be like hunting through a junk yard with a hunch about what a Model T Ford looked like. If the parts should turn up and fit together, the car could be rebuilt, and the hunch about its appearance would be proved correct.

Mangelsdorf and Galinat set to work in their field. In the choice of seed and the care of their crop, these two men were far more painstaking than most conscientious farmers. Each seed was labeled; each stalk of corn had an identification tag; each ear was guarded. To prevent accidental cross-pollination, an experimenter covered the ears with paper bags. When the silks on one plant were ready to receive pollen, he went to another plant and shook pollen from its tassel into a fresh, labeled paper bag. Then he uncovered the ear and capped it again with the new, pollen-filled bag.

Mangelsdorf kept a record of all of his crosses, so he had a pedigree for each new plant. He knew exactly what kind of parents, grandparents, and earlier ancestors it had.

Corn breeders often do this sort of work, but usually they are trying to improve their plants and make them more useful. Mangelsdorf and Galinat planned to develop a corn that would be totally useless—from any practical point of view.

Of course, they had to begin with some idea of what they were aiming for. They had already studied the tiny Bat Cave corn, and they felt sure they knew what its primitive traits were: seeds that popped, glumes covering the seeds, small ears growing high on the stalk, with only a few loose-fitting husks. A plant with these characteristics was what the experimenters wanted to produce.

As a start they chose the kind of pod corn that grows kernels at the base of the tassel and sometimes has no ears at all. This they crossed first with one variety of popcorn, then with another. The results were hybrid pod-popcorn. The hybrids were then crossed, and various traits were combined—again and again and again.

Mangelsdorf and Galinat observed after a while that primitive traits they were looking for seemed to appear together in their pod-popcorn—as if there might be some cause-and-effect relationship between them. The pod-popcorn ear was brittle. It grew near the top of the stalk, and it had only a few husks, which spread open as the kernels ripened. With this fragile construction, free from imprisoning husks, the corn could sow its own seeds, as wild corn did.

The small, high, fragile pod-popcorn ears in Mangelsdorf's garden had still another primitive trait: At the tips of the ears grew a few pollen-bearing male flowers. (In modern corn, male flowers develop in the tassel, which, of course, is completely separated from the ear.)

Now it occurred to Mangelsdorf to take another look at the tiny old cobs from Bat Cave. Sure

*"Wild corn"
recreated by Mangelsdorf
and Galinat (1958)
compared with
modern corn*

enough, he saw what he had overlooked before. At the tips of the cobs were stumps where something had obviously broken off. Little spikes of male flowers must have grown there.

By this time Mangelsdorf and Galinat were sure that their pod-popcorn resembled the ancestral plant in many important ways. Still the ears and kernels were larger than real wild corn would probably have been. Galinat thought he knew the reason. He believed the experimental plants were too well cared for and too well fed. To test this idea, he scattered some pod-popcorn seeds in uncultivated ground against the fence at one edge of the field. Here they had to compete with weeds for sunlight and for nourishment from the soil. As his "wild" corn grew, the effect of competition was soon apparent. The plants were small, just as he had expected. The experimental corn now seemed closer than ever to real wild corn, although the botanists knew it was unlikely that they had completely reconstructed the missing ancestor.

The work of crossing and selecting went on, and the pod-corn gene continued to interest the two men. Mangelsdorf now suspected that the story of this gene was not as simple as he had thought back in 1939. At that time he supposed one drastic change—one mutation—had been responsible for the development of corn with naked seeds. Certainly genetic change of some kind had taken place. But just what was it? Perhaps the gene responsible for glumes was much more complex than it seemed. If so, could it be taken apart and studied? The botanists thought so. Their search into the mysteries of that one tiny speck of hereditary material continued.

Meanwhile another search that had begun earlier was rapidly drawing to a close.

Combined operation

In 1959 botanist Paul Mangelsdorf and archeologist Richard MacNeish mapped an all-out campaign to find not only wild corn but also traces of the ancient people who first learned to grow it. Then, with both people and wild plant tracked down, the two scientists felt sure it would be possible to tell much about civilization in the New World and how it evolved.

Their operation began with just the two of them, but it expanded into collaboration with many specialists. Some would study rocks and soils. Others would analyze various plant remains, or identify animals from their bones, or examine feces to learn what foods people ate thousands of years ago. In addition there would be experts on genetics, on ancient textiles, and on the history of the development of primitive societies. Such collaboration is known as the interdisciplinary approach to a scientific problem, and it is being used increasingly in research. MacNeish was not sure that on this project it would really work, but it seemed worth trying.

The first thing, of course, was to decide where to start looking in the haystack of America for the needle of wild corn. Mangelsdorf could suggest a way to limit the area of search. Botanical evidence, he said, seemed to indicate that wild corn was the kind of grass that grew in high country. Furthermore wild-corn pollen had been found in the Valley of Mexico, which, though a valley, was high above sea level.

Specimens of plant material would have been most likely to survive in a sheltered place in a somewhat arid climate. Mangelsdorf and MacNeish knew from experience that many caves were very dry. So the search might well begin in caves in high, dry country.

Exploration had already been done in the limestone caverns of Guatemala and Honduras and had produced no results. But Santa Marta Cave, in southern Mexico, had yielded some very ancient corn. MacNeish had also found primitive ears in La Perra and other caves in northeastern Mexico. From Mexico City, not quite midway between the two cave areas had come the fossil wild-corn pollen. Perhaps wild corn itself lay buried in some high, dry cave halfway between Santa Marta and La Perra. Perhaps, in this middle area, corn had been domesticated.

Mangelsdorf and MacNeish examined a topographical map of Mexico for areas that were high above sea level. A climatological map told them which parts of the country had a wet climate and which were dry. A geological map showed where the rock was suitable for the formation of caves. High country, dry country, cave country—about halfway between La Perra and Santa Marta—three places in Mexico fitted these requirements.

MacNeish took off on a scouting expedition. The first area he visited did not seem very promising. So he moved on to the next. This one was a long valley running north and south from the city of Tehuacán (tay-wah-KAHN). There might be caves anywhere here. To save time MacNeish sent out a questionnaire to local schoolteachers, asking for leads to the most likely spots. One teacher replied that a man who herded goats knew of a cave in the neighborhood of Tehuacán itself.

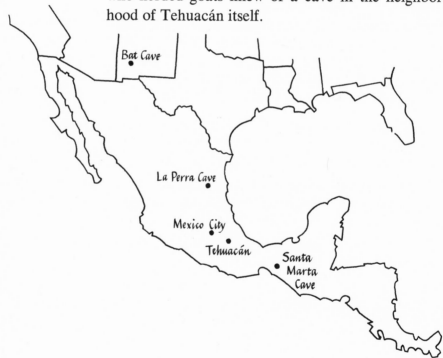

Before long the goatherd was leading the archeologist to a hole in a cliff, then to another and another. MacNeish made test digs in thirty-eight caves
with no luck. In the thirty-ninth, he hit something.
There, toward the end of February, 1960, under
layers of earth and trash, he found the treasure he
sought. When he went north shortly afterward, he
had a present for Mangelsdorf—small corncobs at
least as ancient as the cobs from Bat Cave, which
were about 5600 years old.

Now what MacNeish needed was the money to
pay for a real expedition. Several scientific organiza-
tions gave it, and he found scientists and students
who wanted to help with the work. The Mexican
government lent him a big army truck, which the
diggers promptly named "The Monster."

With shovels, trowels, screens, and brushes, Mac-
Neish's team sifted through the ancient debris in
cave after cave. And the next time MacNeish saw
Mangelsdorf he had a few small grey objects to
show—each hardly bigger than the eraser tip of a
pencil, but without any doubt corncobs. As nearly
as MacNeish could tell, they must have been about
seven thousand years old. If this were so, Mangelsdorf held in his hand the oldest corn ever discovered
anywhere.

Work in 1962 opened up new caves. In one
called San Marcos, in the lowest of many layers of
debris, the archeologists saw at last a number of
cobs which had exactly the characteristics MacNeish

and Mangelsdorf were hoping for:

1 The cobs were tiny—little more than a half inch long—and all nearly the same size. (Wild plants tend to vary in size much less than do cultivated ones.)

2 The cobs were ancient—some as old as 7200 years.

3 The central spike of each cob was very fragile, and the brown and reddish seeds could easily have broken off and sown themselves.

4 The glumes were long, although they did not fully enclose the kernels.

5 There was evidence that a short spike of male, pollen-bearing flowers had grown at the tip of the cob.

6 There was evidence that two husks had grown from around the base of the cob and partially enclosed it. So meager a covering of husks meant that the ears had developed toward the top of the stalk, where leaves were fewer than they were lower down.

7 There were areas near the cave where grasses could have grown wild.

8 There was no evidence in the layer of debris from which these cobs came that the inhabitants of the cave knew anything about farming.

The conclusion seemed almost inescapable: This was wild corn at last! The archeologists took the rest of the day off to celebrate.

Digging continued throughout the valley—in

caves and also out in the open. MacNeish was looking for information about every possible aspect of
ancient Indian culture in this area. He and his crew
turned up remains of other plants the Indians ate,
such as squash and beans. At various levels they
found animal remains, too, and spearheads, tools,
pottery, baskets, woven cloth, ceremonial statuettes
modeled out of clay. There were relics of primitive
huts and of tall pyramids, of campsites, villages,
towns, and cities.

By 1964 the expedition had gathered almost one
million items of archeological interest and nearly
twenty-five thousand specimens of various kinds of
corn. And the work in the valley continued. Almost
fifty different scientists finally came away with material for future books and articles about special as-
pects of what they had found.

Putting all the evidence from all the sciences together, Mangelsdorf, MacNeish, and Galinat could
tell for the first time a story of how corn, and one
part of Indian civilization, originated; how they
affected each other; and how plant and man developed together. Here very briefly is the story:

Perhaps twelve thousand years ago, three or four
bands of people moved into Tehuacán Valley. There
may have been only four or five in each group. The
tools and weapons they brought with them were
made of stone—mainly spears and knives and the
scrapers which they used in preparing skins for use.
This earliest group of wanderers collected wild

*Wild corn from
Tehuacán Valley*

plants and hunted small animals—rats, rabbits, gophers, turtles—though they did manage to kill an occasional wild horse or antelope.

As time passed, each group began to follow regular routes during the year, camping where different plants grew. In the fall they picked nuts. During winter they moved on to places where they could find a cactus that had edible stems. In spring and the wet summertime, they visited spots where such things as squash and wild corn grew.

This wild corn was a small plant that could survive only in a few places exactly suited to it—places where cactus and shrubs did not grow. Its one tiny ear, high up on the stalk, provided only a little nourishment, but hungry men were glad for any food they could get.

Several thousand years went by, and people still lived in small bands, but there were more of them. By 5000 B.C. the valley had perhaps four times as many inhabitants as it had had at first, and they did less hunting and more plant gathering. In the wet season the bands tended to gather together in larger camps at the spots where they found the squash and corn. Then, in fall and winter, each band moved off to make the rounds of its own dry-season camps.

Slowly and by what exact process we do not know, these valley people made a great discovery: If a man—or quite likely a woman—put a seed in the ground it would produce a new plant bearing many new seeds. So agriculture began in the valley.

Now that the groups were not so eternally hungry, they may have had time and energy for new ideas and more elaborate ways of life. Perhaps at this time there appeared medicine men—leaders who were supposed to have special powers over nature and to know something about curing disease. These early priests could not have spent all their time at their special tasks, however. The hunt for food was still too time consuming for anyone to be entirely spared from economic work.

The bands continued to grow in size and number, and they tended to settle down for longer periods in their wet-season camps. First squash and chilies were grown in gardens, then corn, and finally beans. With more food to be had, the valley supported more people—perhaps ten times as many in 3000 B.C. as in the beginning. Because their garden patches required attention, they were not so free to wander as they had been before. Nor did they need to move off so quickly looking for other food. Corn began to change men's habits.

From now on, agriculture really developed. By 2300 B.C. the people of the valley were living together in villages. The population was forty times larger than it had originally been.

While the pattern of life was changing for men, corn was also changing. Because it was cared for, it had less competition from weeds, and so it could grow taller. The cobs were still slender and fragile; the longish glumes almost enclosed the seeds; but the ears were larger and more variable than those of wild corn.

During the next period, farmers began cultivating a distinct new variety of corn. Some of the cobs, Mangelsdorf noticed, had the special horny chaff which he believed was a trait of crosses between corn and teosinte. But there was a problem to consider: Neither teosinte nor Tripsacum grows in Tehuacán Valley today, and, so far as the archeological evidence showed, those plants never did grow there. Two explanations were possible: (1) Mangelsdorf might have been wrong about the meaning of the horny chaff: it might not indicate crossing with teosinte. Or (2) this new variety of corn had been imported into the valley from some other place where the maize plant could have crossed with teosinte.

The archeologists and botanists agreed that the second explanation was the more probable. In another valley not far from Tehuacán, both teosinte and Tripsacum do grow. And, what's more, there was definite indication that, during the period when this new variety showed up in Tehuacán Valley, the people there were trading with other groups in the surrounding territory.

This and other new varieties of corn had bigger ears, shorter glumes, and more kernels, and could provide more food. By 200 B.C. the population was one hundred fifty times larger than it had been to begin with. People had developed the arts of pottery, woodworking, and weaving. They were full-time farmers, and they had learned to irrigate their fields. The corn they grew was even more productive than

Ancient artifacts found in Tehuacán Valley

before; and some of it began to resemble varieties which grow in Mexico today. Still these farmers, and the more advanced people who followed them, continued to collect wild corn to eat along with their improved varieties. MacNeish discovered tiny wild corncobs that had been left in caves as late as 600 A.D. But neither he nor anyone else has found wild corn that grew at a later date.

What, then, happened to the wild maize plant? Perhaps men had gradually dug it all up when they made their fields in the places where it grew best. There was also another possibility. Perhaps wind had blown pollen from fields of improved corn to the remaining patches of wild corn. Offspring of the wild parent and the field-grown parent would have been less able to seed themselves, and the plants would eventually have died out. If this did indeed happen, then domesticated corn killed its wild ancestor.

Irrigation and other improvements in farming meant still more food. When the Spaniards arrived in the valley in 1536 the population was five thousand times its original size. Between 200 B.C. and the Spanish Conquest, people had been creating a civilization in the valley. Most of them remained laborers in the fields, but there was enough food to free some men for other work or for leisure.

Pyramids and ceremonial cities were built. Some of them covered whole mountaintops. A salt industry was founded, artists and craftsmen flourished, and commerce spread. Apparently the valley was now divided into a number of little kingdoms. In each of these lived separate classes of priests and lords, fed by the corn farmers. The ceremonial cities were probably ruled by priest-kings, with the help of officials and specialists who supervised the building and the irrigation systems.

It took a long time, perhaps twelve thousand years, but the Indians who started out in tiny bands, gathering seeds and hunting small creatures for bare survival, developed an advanced culture in Tehuacán Valley. And they were able to do so because they found corn and learned to raise it, and because corn evolved under their care from a small obscure weed into the plant which Edward East called "the prince of grasses."

Collaboration—the interdisciplinary approach—had brought to a quick and dramatic conclusion the century-old search for the origin of corn. Here was a great victory for science, won by combined operations. It was also a victory for Paul Mangelsdorf and Richard MacNeish, the two imaginative men who had planned the final campaign to solve the mystery of maize. By prying into the story of one kind of grass, they had opened up the story of civilization in the Americas. What had begun with a close look at maize ended with a clearer vision of New-World man than anyone ever had before.

Where next?

The work at Tehuacán gave one specific time and place for the domestication of corn. Meanwhile another investigation had been going forward, and it produced a more sweeping answer to the question of corn's geographical origin.

For many years Barbara McClintock, geneticist and authority on the maize plant, had been intrigued by microscopic dark spots or knobs which often appeared on the chromosomes in the cells of corn and its relatives. Neither she nor anybody else could explain what function, if any, these knobs had. The most that could be said at first was that they fell into different patterns of size and position on the chromosomes. In some varieties of corn the knobs were large and numerous; in others they were scarce and small, almost invisible; in still others they might be somewhere in between the extremes.

As more and more thousands of specimens were examined, it became increasingly clear that certain patterns were concentrated in certain geographical areas. There was a definite correlation between points on the map and knobs on chromosomes of modern corn. But could these same knobs be used as tools for tracing back different varieties of corn to their place, or places, of origin?

When Mangelsdorf and Reeves did their experiments in Texas, they studied the knob chromosomes of corn and its relatives, and they found evidence which seemed at the time to support their tripartite theory. Barbara McClintock and some other botanists disagreed with that interpretation.

Studies of the knobs went on, and one thing became certain. Knob patterns in the races of modern corn indicate that wild corn of several different varieties must have grown in the Americas. It may have been domesticated by more than one group of people in more than one place, perhaps at different times.

Quite possibly a knob expert will someday be able to trace the routes by which each kind migrated and crossed with other kinds, until all the varieties of modern corn evolved. Meantime archeologists are examining new sites, looking for more clues to corn's wild ancestry.

Questioning and search into the secrets of the pod-corn gene also go on. So do intense investigations of corn's relatives, Tripsacum and teosinte. Much detective work has led Walton Galinat to a new theory and a revised picture of corn's family tree. Tripsacum, he now believes, is really a hybrid plant—a cross between wild corn and a grass called Manisuris. This new approach may help to account for the tremendous vigor of modern corn. It may also help

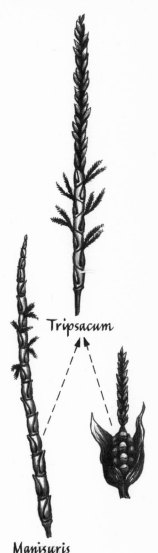

Tripsacum

Manisuris

*Seneca Indian
ceremonial mask
woven from corn husks*

to explain how the plant could have evolved in so few thousand years.

All of these investigations have increased man's understanding of corn's past. And understanding can lead to further improvement of what is already the most useful plant in the world.

From its beginning the story of maize has been a story of increasing usefulness. At first the Indians grew it almost exclusively as food. People ate the kernels and usually threw away the rest of the plant. A few tribes did learn to weave husks into mats, sandals, baskets, and ceremonial masks. In Central America the stout stalks of one tall variety have been used to make fences and walls of huts. But, by and large, Indians raised corn to feed themselves.

When Europeans arrived in the Americas, they soon expanded the plant's usefulness—not through any special genius the white man possessed, but because of his animals. He imported with him cattle, horses, pigs, sheep, goats—domesticated animals which the Indians lacked and which thrive on corn. A cow makes much better use of corn than does a man, for it eats the whole plant and finds nourishment in the sweet pulpy stalks and lush leaves, as well as in the ears. Men eat beef and indirectly benefit from the whole corn plant.

The Indians tended to grow those varieties which are most palatable to human beings. European settlers began to experiment with varieties which were especially nourishing for their less finicky beasts. They were particularly pleased with crosses between a tough New England corn called *flint* and a many-eared southeastern kind called *gourd-seed*. The result was a highly-productive *dent* corn, so called because, when the kernels dry, the plump top of each one shrinks to a dimple or dent.

Everywhere on the New World continents farmers discovered the benefits of the marvellous variation in corn. There was maize for many climates and territories—for the rocky soil of New England, the deserts of the Southwest, the short summers of Canada, the peaks of the Andes. In the nineteenth century, when men moved into the midwestern United States with plows and oxen and seed to plant, they bred corn mainly as feed for animals. The uniform, yellow ears of their unappetizing cattle corn were quite different from anything the Indians or the first white settlers had grown—and much more fruitful.

The work of corn improvement continued. There were no huge seed companies then, as there are now, to sell standard varieties of scientifically bred corn seed to farmers. It was each man for himself. As a rule a farmer had to develop his own seed, and what he lacked in scientific training he tried to make up in common sense. After the harvest he selected his best ears for next year's planting. One well-known improver got a reputation for his exceeding

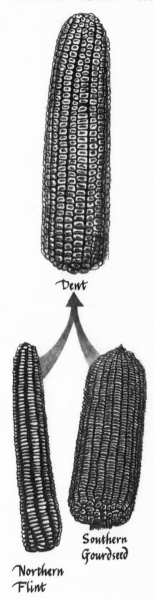

Dent

Northern
Flint

Southern
Gourdseed

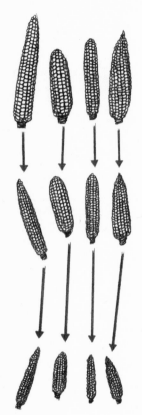

When corn is inbred, it becomes smaller and weaker.

caution. To protect his seed corn through the winter he stored it between the mattresses on which he slept.

Farmers had their own notions about what the best corn was, and they selected their seeds for the traits they liked. Some fancied a plant with one or two big ears. Others preferred stalks that had more, though smaller, ears. Around the turn of the twentieth century the Midwest was swept by an enthusiasm for "pretty" corn—and this was defined as a plant with a large stocky ear which had eighteen to twenty-four straight, regular rows. Farmers vied with each other to produce the most uniform batches of it.

During the next twenty-five years, just when farmers had come to a general agreement on what the best corn was, agricultural scientists developed new varieties which were to change farmers' notions radically. These were the hybrid corns; and it was a midwestern chemical engineer, Edward Murray East, who did much of the pioneering in their creation.

East began as consultant in an experiment at the University of Illinois where scientists were hoping to improve the chemical content of corn. They wanted more protein and less starch in the kernels, for the better nourishment of cattle and men. Before long it was clear to East that genetics rather than chemistry held the key to new varieties. His work continued at the Connecticut Experiment Station. There he developed the ideas about hybrid corn which were finally perfected and made practical by another midwesterner, Donald F. Jones.

At first it seemed doubtful that American farmers could be won over to the new hybrid varieties. For one thing, a batch of hybrid seed was good for only a single huge crop. A fresh batch had to be bought from a commercial seed producer every year. To many farmers that seemed wasteful and extravagant. Besides, a man did not easily give up the habit of using the home-grown seed that was in a sense his own individual creation. Nevertheless by the end of the Second World War most of the corn grown in the United States was hybird corn, and farmers in other nations began learning to develop their own hybrids.

Hybrid corn is the most opulent of all man's plants. No other crop in the United States covers so many acres of land or produces more wealth. No other grain plant returns so much for so little. One seed can grow into a stalk that bears an ear with as many as 1200 new kernels. The American standard of living greatly depends upon corn, and the spread of improved hybrid-corn agriculture can help to make a better life for the people of most of the world.

The development of hybrids was, however, only one of the contributions of twentieth-century science to the increasing usefulness of the plant. The products of maize have found their way into a multi-

When inbred strains of corn are crossed, they produce offspring which have great vigor. Commercial hybrid corn is the result of a double cross: Four strains of inbreds are crossed to produce two hybrids, which are then crossed again to produce seed for planting.

tude of industrial products we use every day. Corn is often an essential ingredient of:

antibiotics	pencil erasers
aspirin pills	sandpaper
baking powder	dry-cell batteries
beer	paper
ice cream	books and bookbinding
ice-cream cones	crayons
cosmetics	chalk
soap	detergents
catsup	inks and dyes
chewing gum	explosives
licorice	oilcloth
marshmallows	paint and paint remover
peanut butter	photographic film
pickles	window shades
vinegar	matches
margarine	plastics
textiles	drinking straws

And then, of course, there are corncob pipes.

Corn comes to all these products—and thousands more—through the work of the corn-refining industry, which operates rather like the petroleum-refining industry. Both crude petroleum and the kernel of corn are made of a number of different chemical substances. Refining is the process of separating these substances, purifying them, and sometimes changing them so that each can be used for its special properties. Gasoline, butane gas, and cheap candle wax come from the same black liquid. Yellow salad oil, smooth starch, white sugar, and transparent film come from the same hard kernel of maize.

One out of every three bushels of corn produced in America ends up in a refinery—three hundred railway cars full every day of the year. The refining process is complicated and ingenious, and nothing goes to waste. Oil, syrup, sugar, and starch are the basic corn products. Each one of these comes in different varieties and has many uses. Once the corn is treated and these products extracted, there are some leftovers—the skin of the kernel (or hull) and the water in which the corn is soaked for refining. This soak-water is rich in nourishing corn chemicals, and it is combined with the hulls to make food for animals.

The corn-products industry has only begun to exploit all the possibilites for the use of this versatile plant. One company alone has two hundred scientists at work developing new ideas. Perhaps, for example, automobiles may some day be put together with an especially strong corn adhesive instead of rivets and bolts. Because the chemical makeup of cornstarch is somewhat similar to that of petroleum, maize may be turned into automobile fuel.

One of two closely related substances in cornstarch can be made into a film resembling cellophane, but stronger and different in important ways. Unlike cellophane, it dissolves in water, and it is edible.

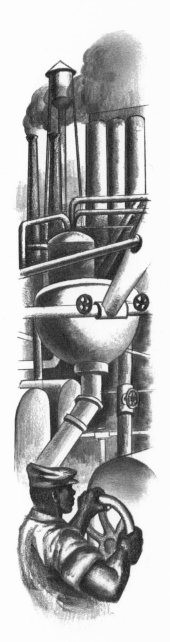

Someday frozen vegetables may come wrapped in this film. The whole package can simply be dropped into boiling water, and the wrapping will disappear. Unfortunately the film substance is hard to extract from cornstarch, and ordinary varieties do not contain a great deal of it. This makes it expensive.

There have been many such problems which corn scientists have learned to solve. Sometimes they simply watch and wait. Those who want to make the edible film have already tested many varieties of corn, looking for a kind that may produce more than the usual amount of the right material. They have found one, but it isn't quite productive enough. Starting with this plant, a hybrid-corn breeder may be able to cross and select until he has developed a new variety even richer in the film substance. The maize plant is so variable that a breeder can do almost anything he likes with it. If industrial chemists tell the corn geneticist what they want, there is a good chance that eventually he will be able to breed a variety that exactly suits their needs.

One special kind of corn may help to solve a problem for space travelers. How can a spacecraft carry enough oxygen for astronauts to breathe, for instance, during a long stay on the moon? A partial answer seems to be to take plants along. People breathe in oxygen and breathe out carbon dioxide. Plants work the other way around—they take in carbon dioxide and give off oxygen. Ordinary corn does just that. But at the Connecticut Agricultural

Experiment Station, where Edward East worked on hybrid corn and young Paul Mangelsdorf began his career, scientists have discovered a variety which is particularly good at returning oxygen to the atmosphere.

Corn may be the first plant on the moon.

Acknowledgments

We should like to acknowledge the kind assistance of a number of men and women who made this book possible. Some of their names appear in our story, for it is their story, too. Others worked behind the scenes. All offered valuable help.

Dr. Paul C. Mangelsdorf talked with us, interrupted his own writing to read drafts of our manuscript, and cheerfully tried to steer us away from mistakes and misinterpretations. He was pleased to have us tell as we saw fit the story in which he has long been a controversial character. If errors remain, they are our responsibility, not his.

Dr. Mangelsdorf and Dr. Walton C. Galinat toured us through their experimental cornfield one summer day, answered innumerable questions, and told us where to find more answers. In addition, Dr. Galinat took time out from his work at the Waltham Field Station, University of Massachusetts, to check Frank Cieciorka's illustrations.

Dr. Richard S. MacNeish found time between sessions at a Society for American Archaeology meeting to answer questions and add details to the story.

Dr. Frederick A. Peterson, Assistant Director of the Tehuacán Project, hospitably offered a jeep ride among modern Mexican cornfields, a visit to one of Tehuacán Valley's ancient caves, and much enlightening talk.

Dr. George F. Sprague gave us a long afternoon in his field at the Agricultural Research Center, Beltsville,

Maryland, explaining his work and sharing our delight, if not our amazement, in his collection of most unusual varieties of corn. Miss Alice Robert, his assistant, added to the story of corn research and how it was being done.

At Cold Spring Harbor, Long Island, Dr. Barbara McClintock gave us good advice as well as helpful material about her investigations of corn chromosomes.

At the Missouri Botanical Garden in St. Louis, Dr. Hugh C. Cutler, surrounded by cartons full of ancient plant specimens, told of continuing botanical-archeological work.

We are indebted to Mr. Leon Svirsky for introducing us to the mystery of corn and for the many hours he spent helping us to work our way through it. To Dr. Carroll Lane Fenton and to Miss Marian H. Scott go thanks for reading a draft of the manuscript and for making valuable suggestions.

Our special thanks to Frank Cieciorka, an illustrator who does not know how to draw indifferently. He set out to make this his book as well as ours—to make the corn plant as surprising and understandable to the eye as we tried to make it to the mind.

Finally, our gratitude to Franklin Folsom whose patient scrutiny and encouragement were most responsible for shaping our material into a story we like.

Mary Elting
Michael Folsom

For further reading

Edgar Anderson, *Plants, Man and Life*. Little, Brown, 1952.

Vance Bourjaily, "Corn of Coxcatlán," in *Horizon*, Spring, 1966.

Robert Claiborne, "Digging up Prehistoric America," in *Harper's Magazine*, April, 1966.

A. Richard Crabb, *The Hybrid Corn Makers*. Rutgers University Press, 1947.

Anne Ophelia T. Dowden, *Look at a Flower*. Thomas Y. Crowell, 1963.

Gordon F. Ekholm, "Transpacific Contacts," in *Prehistoric Man in the New World*, edited by Jesse D. Jennings and Edward Norbeck. University of Chicago Press, 1964.

Jack R. Harlan, *Plant Scientists and What They Do*. Watts, 1964.

Richard S. MacNeish, *First Annual Report of the Tehuacán Archaeological-Botanical Project*. Robert S. Peabody Foundation for Archaeology, Andover, Mass., 1961.

Richard S. MacNeish, *Second Annual Report of the Tehuacán Archaeological-Botanical Project*. Robert S. Peabody Foundation for Archaeology, Andover, Mass., 1962.

Richard S. MacNeish, "The Origins of New World Civilization," in *Scientific American*, November, 1964.

Paul C. Mangelsdorf, "The Mystery of Corn," in *Scientific American*, July, 1950.

Paul C. Mangelsdorf, "Wheat," in *Scientific American,* July, 1953.

Paul C. Mangelsdorf, Richard S. MacNeish, Gordon R. Willey, "Origins of Middle American Agriculture," in *Handbook of Middle American Indians,* Vol. 1, R. Wauchope, editor. University of Texas Press, 1964.

Howard T. Walden 2nd, *Native Inheritance.* Harper & Row, 1966.

Henry A. Wallace and William L. Brown, *Corn and Its Early Fathers.* Michigan State University Press, 1956.

Paul Weatherwax, *Indian Corn in Old America.* Macmillan, 1954.

The authors found much information in the above books, articles, and reports. They also consulted many other publications, among which the following were especially useful. Readers who want to adventure further into the story of corn may do so with the aid of these more technical scientific works.

Botanical Museum Leaflets, Harvard University, Vol. 13: 213-247; Vol. 14: 157-180; Vol. 16: 229-240; Vol. 17: 101-178; Vol. 18: 329-440; Vol. 19: 163-181.

Alphonse de Candolle, *Origin of Cultivated Plants.* Hafner (Reprint), 1959.

Herbert W. Dick, *Bat Cave.* School of American Research, Santa Fe, New Mexico, Monograph 27, 1965.

Herbert W. Dick, "The Bat Cave Pod Corn Complex," in *El Palacio,* May, 1954.

Theodosius Dobzhansky, *Evolution, Genetics and Man.* Wiley, 1955.

Walton C. Galinat, "The Evolution of Corn and Culture in North America," in *Economic Botany,* October-December, 1965.

Richard S. MacNeish, "Ancient Mesoamerican Civilization," in *Science,* 7 February, 1964.

Paul C. Mangelsdorf, "Ancestor of Corn," in *Science,* 28 November, 1958.

Paul C. Mangelsdorf, Richard S. MacNeish and Walton C. Galinat, "Domestication of Corn," in *Science,* 7 February, 1964.

Paul C. Mangelsdorf and Robert G. Reeves, *The Origin of Indian Corn and Its Relatives.* Texas Agricultural Experiment Station Bulletin, 574, 1939.

H. F. Roberts, *Plant Hybridization Before Mendel.* Princeton University Press, 1929.

George F. Sprague, editor, *Corn and Corn Improvement.* Academic Press, 1955.

Paul Weatherwax, *The Story of the Maize Plant.* University of Chicago Press, 1923.

FILMS

Corn and the Origins of Settled Life in Meso-America, a 16mm sound color film in two parts, produced and distributed by Educational Services, Inc., 55 Chapel Street, Newton, Mass.

The Great Story of Corn, a 16mm sound color film, produced and distributed by Funk Bros. Seed Company, Bloomington, Ill.

Index

Agricultural Experiment Station(s), 34, 35, 36, 37, 38, 42, 43, 108, 109
archeology, 64, 71-75, 88-99
Ascherson, Paul, 14
Azara, 47

Barghoorn, Elso, 80
Bat Cave, 64, 65-70, 75, 81, 84, 91
Bonafous, Mathieu, 54
Bradford, William, 53
Burbank, Luther, 16-18, 19, 22, 23

Candolle, Alphonse de, 7, 12, 25, 32, 55, 62
carbon-14 dating, 70
cave(s), 63, 89-91
  Bat Cave, 64, 65-70, 75, 81, 84, 91
  La Perra Cave, 71-75, 89
  San Marcos Cave, 91
  Santa Marta Cave, 89
chromosome(s), 21, 100-101
Clisby, Kathryn, 80
Columbus, Christopher, 50, 51, 53, 55, 59

corn
  archeological, 63, 64, 65-70, 71-75, 88-99
  ceramic representation of, 57, 76-77
  colors, 4, 19, 21
  coyote, 14, 15
  crossed with teosinte, 16-18, 48, 49, 68-69, 97
  crossed with Tripsacum, 37-42, 43, 44, 49, 61
  dent, 103
  ear structure, 29
  evolution of, 3-9, 15-18, 19, 26-32, 48, 68-69, 97
  family tree, 28, 101
  flint, 103
  fossil, 75, 76-82
  geographical origin of, 50-60, 62, 81, 100-101
  germination, 26
  gourd seed, 103
  hybrid, 34, 35, 36, 37, 105
  hybrid vigor, 34
  inbred, 34
  improvement, 103
  legends, 10-12
  origin of name, 53